Pests and Diseases in Spices, Plantation and Tuber Crops

Pests and Diseases in Spices, Plantation and Tuber Crops

Dr. S. Parthasarathy
Assistant Professor (Plant Pathology)
Amrita School of Agricultural Sciences
Amrita Vishwa Vidyapeetham, Coimbatore

Dr. R. Kalaivanan
Assistant Professor (Plant Pathology)
Vanavarayar Institute of Agriculture, Pollachi
(Affiliated to Tamil Nadu Agricultural University)
Coimbatore

Dr. R. Sangavi
Assistant Professor (Entomology)
Vanavarayar Institute of Agriculture, Pollachi
(Affiliated to Tamil Nadu Agricultural University)
Coimbatore

Dr. P. Lakshmidevi
Teaching Assistant (Plant Pathology)
Institute of Agriculture, Kumulur, Trichy
Tamil Nadu Agricultural University, Coimbatore

-EPH-
Elite Publishing House

CRC Press
Taylor & Francis Group
Boca Raton London New York

CRC Press is an imprint of the
Taylor & Francis Group, an **informa** business

First published 2025
by CRC Press
4 Park Square, Milton Park, Abingdon, Oxon, OX14 4RN

and by CRC Press
2385 NW Executive Center Drive, Suite 320, Boca Raton FL 33431

CRC Press is an imprint of Informa UK Limited

British Library Cataloguing-in-Publication Data
A catalogue record for this book is available from the British Library

ISBN13: 9781041034278 (hbk)
ISBN13: 9781041034285 (pbk)
ISBN13: 9781003623779 (ebk)

DOI: 10.1201/9781003623779

Typeset in A Caslon Pro
by Elite Publishing House, Delhi

-EPH-

Contents

About the Authors

Dr. S. Parthasarathy currently serves as an Assistant Professor in Plant Pathology at the Department of Plant Pathology at Amrita School of Agricultural Sciences. He earned his PG and Ph.D. degrees from Tamil Nadu Agricultural University, Coimbatore. Dr. Parthasarathy has been honored with various prestigious fellowships, including the BIRAC Post Master's Innovation Fellow in DBT (2015–17), the Canadian International Development Research Centre-Nano Project JRF (2013–14), the UGC Non-Special Assistance Programme (2012), and the Ministry of Human Resource Development Fellowship during his undergraduate studies in 2009. He has also received medals during his undergraduate program and numerous meritorious awards at both international and national conferences for his research contributions. An emerging expert in Plant Pathology, he has authored 55 research articles in peer-reviewed journals, written 16 books, and contributed approximately 45 chapters to various books of national and international acclaim. Furthermore, he has served as a member of the editorial boards of national journals and various international scientific societies. His areas of specialization encompass Biological Control and Molecular Plant Pathology.

Dr. R. Kalaivanan holds the position of Assistant Professor specializing in Plant Pathology at the Department of Crop Protection at Vanavarayar Institute of Agriculture in Pollachi. He obtained his postgraduate and Ph.D. degrees from the Agricultural College & Research Institute in Madurai (TNAU). As a young and accomplished Plant Pathologist, he boasts an impressive record, including the publication of 2 international research papers, 5 national research papers, a book, and contributions to 4 book chapters. He actively participates in several scientific societies and has a history of engagement in national and international conferences, earning multiple awards. Moreover, he is actively involved in extension activities, particularly in providing hands-on training to farmers to address real-world issues encountered in the field. His areas of expertise encompass Biological Control, Plant Bacteriology, and Molecular Plant Pathology.

Dr. R. Sangavi holds the position of Assistant Professor in Entomology at the Department of Crop Protection at Vanavarayar Institute of Agriculture, Pollachi. She completed her Ph.D. from Navsari Agricultural University, Gujarat. She is a recipient of the ICAR SRF fellowship and has authored 12 research papers, 5 book chapters, and 2 books. She also serves as a reviewer for two national journals and has organized several national and international conferences. Dr. Sangavi has represented as a resource person in various farmer training programs and has served as a research guide for students and a member of numerous agricultural projects. Her area of specialization includes Biological control, Biopesticides, and Acarology.

Dr. P. Lakshmidevi holds the position of Teaching Assistant in Plant Pathology at the Institute of Agriculture, Tamil Nadu Agricultural University. She completed her Ph.D. at Tamil Nadu Agricultural University, Coimbatore. She is a recipient of RGNF fellowships during 2011-14, a Young Plant Pathologist Award, and the Best Book Contributor Award in 2022. Dr. Lakshmidevi has conducted 2 Farm Testing trials, 4 Front Line Demonstrations, two farmers' field schools, and 43 on and off-campus training sessions. She has published 4 books with ISBNs, 12 research articles, 29 conference papers, and 23 book chapters with ISBNs.

1 Introduction

Horticulture is a science, as well as, an art of production, utilization and improvement of horticultural crops, such as fruits and vegetables, spices and condiments, ornamental, plantation, medicinal and aromatic plants. Horticultural crops require intense care in planting, carrying out intercultural operations, manipulation of growth, harvesting, packaging, marketing, storage and processing. India is the second largest producer of fruits and vegetables in the world after China. In India, about 55–60 per cent of the total population depends on agriculture and allied activities. Horticultural crops constitute a significant portion of the total agricultural produce in India. They cover a wide cultivation area and contribute about 28 per cent of the Gross Domestic Product (GDP). These crops account for 37 per cent of the total exports of agricultural commodities from India.

India is the largest producer, consumer and exporter of spices and spice products in the world. Over 100 plant species are known to yield spices and spice products, among which about 50 are grown in India. Pepper is considered to be the 'king of spices,' and cardamom the 'queen of spices'. In India, pepper covers 1,25,000 ha with an annual production of 53,000 tonnes. Kerala, Karnataka and Tamil Nadu produce more than 90 per cent pepper. Cardamom (small) is also grown in these three states of India over 92,000 hectares of land, with a total production of 18,000 tonnes in a year. In India, large cardamom is mainly cultivated in Sikkim and West Bengal over an area of 30.039 ha producing 5401 tonnes of cardamoms. Ginger is grown in almost all the states and the total acreage of this crop is about 1.36 lakh ha and the annual production is 6.83 lakh tonnes of dry ginger. Turmeric is grown in many states and covers an area of 1,94,000 hectares with an annual production of 9,71,400 tonnes. Chilli occupies an area of 7.94 lakh hectars in India with an average annual production of 13.04 lakh tonnes. India produces half the production of chilli in the world.

Since spices come from different kinds of plants, belonging to a number of families, the pests that have adapted themselves to feeding on these aromatic plants are also of different types. Apart from the field pests of the spice plants, a number of insect pests of stored products also damage spices in storage. The crop loss is estimated to be approximately 20-40% of production due to various pests, diseases and weeds. Being an export oriented spice and plantation crops, the responsibility of producers/growers is to be high. The spice production sustains losses for a number of reasons. Diseases and pests frequently attacking spice crops are considered as a major factors affecting their production. Numerous pest and pathogens species impair the production of high-value spice crops such as turmeric, black pepper, ginger, nutmeg, cardamom, tamarind, vanilla, and clove, and may suppress their yields up to 21%.

Plantation crops refer to crops that are grown on a large scale in a contiguous area and are owned and managed by an individual or a company. Tea, coffee, rubber, cocoa, coconut, arecanut, oil palm, palmyrah, and cashew are among the crops grown. These are high-value commercial crops with greater economic importance, and they play an important role in improving the Indian economy, particularly in terms of export potential, job creation, and poverty alleviation, particularly in the rural sector. The Ministry of Agriculture is in charge of coconut, cashewnut, cocoa, arecanut, oil palm, and palmyrah, while the Ministry of Commerce is in charge of tea, coffee, and rubber. The three southern states, Karnataka, Tamil Nadu, and Kerala, are home to coffee and tea gardens, rubber plantations, and spice crops, generating employment to more than 13 lakh people. Almost 40% of tea produced in the southern States is exported and 31% of pepper grown here, too, goes to other countries. However, the plantation sector reels under issues such as challenges of pests and diseases, the need to improve yield, and necessity of growers to focus on value addition.

Tuber crops attain importance as the second most important energy provider crops after cereals. The major tuber crops of India are cassava, sweet potato, yams, elephant foot yam, taro, tannia, yam bean, arrow root etc. They played a key role in the diet of early human population before evolution of settled agriculture. Root and Tuber Crops are the most important food crops after cereals. They have the highest rate of dry matter production per day and are major calorie contributors. Tuber crops find an important place in the dietary habits of small and marginal farmers especially in the food security of tribal population. Tuber crops not only enrich the diet of the people but also possess medicinal properties to cure many ailments or check their incidence. Tuber crops constitute a major component of food for indigenous and tribal people throughout the world. Though yams and taro belongs to India, cassava was introduced from South America by the Portuguese during 16th and 17th century only. This outsider was quickly accepted by the people and it became a symbol of food and nutritional security of the rural India.

Among the tuber crops, Cassava is the most important one in the tropics and it ranks fourth, after rice, sugarcane and maize, as a source of calories for human consumption. It is a major carbohydrate food for about 500 m people in the world, and in Africa, it is the most important source of calories in the human diet. Cassava, sweet potatoes and elephant foot yam are the most important tuber crops in India due to its large area under commercial cultivation and its varied uses. Cassava is grown in an area of 0.23 million ha^{-1} producing 4.65 million tons with an average productivity of 20.39 t ha^{-1} during 2018. Tamil Nadu, Kerala and Andhra Pradesh are the major cassava growing states in India, while sweet potatoes are mainly grown in Odisha, Kerala and West Bengal. While there is widespread consensus that insect pests and diseases can play an important role in causing food insecurity, there is little information about the magnitude of yield and post-harvest losses of tuber crops in India,

Many chemical and non-chemical control options are available to spices, condirments and plantation crops for dealing with insect pest and diseases problems, including the modification of cultural practices, use of resistant varieties, application of synthetic insecticides, application of botanicals and conservation practices for natural enemies/ bioagents of the insect pests and diseases. Integrated pest and disease management (IPDM) simply signifies the application of the best mix of environmentally sound techniques in order to keep pests below numbers that can cause unacceptable damage. In practice, IPDM means avoiding the dependence on one single method for controlling pests and diseases; various measures are undertaken, reinforcing one another, to ensure that pest populations remain below that which will cause unacceptable damage. The ultimate objective is to reduce the dependence on chemical control through the use of synthetic pesticides as the main input for crop pest and disease management. Reduction in the use of pesticides lessens environmental pollution and minimizes the risks of poisoning the user and consumer alike.

References

Pruthi, J. S. (1993). Major spices of India. Crop management and post-harvest technology. Major spices of India. Crop management and post-harvest technology.

Rathore, M. S., & Shekhawat, N. S. (2008). Incredible spices of India: from traditions to cuisine. American-Eurasian Journal of Botany, 1(3), 85-89.

Krishnaswamy, K. (2008). Traditional Indian spices and their health significance. Asia Pacific journal of clinical nutrition, 17(S1), 265-268.

Peter, K. V. (Ed.). (2006). Handbook of herbs and spices: volume 3. Woodhead publishing.

Shukla, A., & Yadav, N. (2018). Role of Indian spices in Indian history. International Journal of Management Re Gayathri, R., & Saravanan, M. (2014). A Study on the Performance of major spices in India. Asia Pacific Journal of Management & Entrepreneurship Research, 3(1), 37. Research and Reviews, 8(11), 1.

Hebbar, K. B., Balasimha, D., & Thomas, G. V. (2013). Plantation crops response to climate change: coconut perspective. Climate-resilient horticulture: adaptation and mitigation strategies, 177-187.

Hebbar, K. B., Berwal, M. K., & Chaturvedi, V. K. (2016). Plantation crops: climatic risks and adaptation strategies. Indian Journal of Plant Physiology, 21, 428-436.

Kumar, N. (1997). Introduction to spices, plantation crops, medicinal and aromatic plants. Oxford and IBH Publishing.

Shanmugavelu, K. G., Kumar, N., & Peter, K. V. (2002). Production technology of spices and plantation crops. Agrobios.

Hebbar, K. B., Pandiselvam, R., Beegum, P. P., Ramesh, S. V., Manikantan, M. R., & Mathew, A. C. (2022). Seventy-five years of research in processing and product development in plantation crops-Coconut, arecanut and cocoa. International Journal of Innovative Horticulture, 11(1), 103-119.

2 Arecanut

INSECT PESTS

Spindle bug, *Mircarvalhoia arecae*

Damage symptoms

- Adults and juveniles alike consume sap from closed spindles.
- Linear dark brown necrotic lesions caused by eating appear on spindles or exposed leaves.
- After some time, the necrotic tissue begins to pull away, exposing the underlying tissue.
- Affected spindles don't always open all the way.
- When infection levels are high, the leaves are shredded.

Pest identification

- The adults are red and black, while the juveniles are a shady green.

Management

- To replace the several phases of the beetle, dig and fork the soil before and after the monsoon.
- Apply a spray containing 0.1% Carbaryl to the crowns. The axils of the leaves should be covered by the spray. If pest problems persist after 30–35 days, reapply the pesticide.
- Keeping Phorate/Carbofuran granules in perforated polythene sachets at the leaf axil of areca palms to prevent *M. arecae* from entering the crown.

- Thiamethoxam 25 WG (2 g) in a perforated poly-sachet is inserted every three months into the axils of the first two leaves. This method is helpful for controlling especially pervasive pest problems in outdoor spaces.
- If pests return, spray Dimethoate (Rogor 30 EC) at a rate of 1.5 ml per litre of water and repeat the process every 15 to 20 days. It's important for the spray to penetrate deep into the leaf axils.

Inflorescence caterpillar, *Tirathaba mundella*

Damage symptoms

- Its diet consists primarily of rachillae and young female flowers.
- The first signs are the presence of bore holes on the spathe and the resulting damage to the young racillae and male flowers.
- In their later stages, caterpillars create silken webs at the tips of the inflorescences, after having bored through the sensitive button.
- Frass is also seen to be extruding through the webs.
- The inflorescence is characterised by delayed spathe opening.
- These can even bore through sensitive buttons and nuts in extreme circumstances.

Pest identification

- Adult moths have a wingspan of 28-35 mm with a greyish upper wing surface and a cream colour underside.

Management

- If the spathe of the inflorescence has holes, you should cut it open and remove the inflorescence in its entirety. If only a little section has been damaged, you can spray it with 0.05% Monocrotophos and replace it.
- This caterpillar can be effectively sprayed with Malathion 50 EC at a concentration of 2 mg per litre of water.

Scale insect, *Aonidiella orientalis*

Damage symptoms

- Spread by settling on plant parts like leaves, spathe, leaf sheaths, bunches, and nuts, from which they can extract plant sap.
- Between the months of October and February, a huge influx of people is

seen.

» Premature flower and button drop, as well as nut yellowing, are all effects of overfeeding.

» If the infestation is bad enough, the kernel might not develop at all, turning black and shrivelling up instead.

Management

» Shade the garden as needed.

» Neem oil spray, 5 ml per litre of water.

» Diazinon, dichlorvos (DDVP), phosphamidon, monocrotophos, and quinalphos sprayed at 0.04% are all effective pesticides.

» At a rate of four to five beetles per palm tree, *Chilocorus nigritus* should be released on a regular basis.

» Protect natural enemies like the coccinellid beetles (*C. nigritus* and *C. circumdatus*).

Root or white grub, *Leucopholis lepidophora*

Damage symptoms

» Damage to areca palms occurs during the grub stage, when the insects feed on the tree's roots.

» This causes the palm's leaves to turn yellow, its yield to decrease, its stem to thin, and its overall appearance to be unhealthy.

» Injuries to the palm's roots have made it unstable.

» If the seedlings have progressed too far, they can be readily uprooted.

» The beetle first appears in the dark of night following the pre-monsoon rains.

Pest identification

» Seven to eight months is the duration of the grub stage, which consists of three instars.

» Adult beetles tend to be a dark chestnut brown.

Favorable conditions

» Extremely common in places where water cannot drain freely.

Management

- Make sure the garden has proper drainage.
- Adult beetles are most easily captured and killed between 6:30 and 7:30 p.m., right before the monsoon season begins.
- In the months of May and June, and again in the months of September and October, you should dig and fork the soil around the palm's basin to a depth of 10 to 15 centimetres.
- At the start of the south-west and north-east monsoon seasons, it is best to apply organic amendments like Neem cake at a rate of 2 kg per palm tree every year. By following these steps, fresh root growth can be encouraged.
- In the months of May and June, and again in the months of September and October, spray the palm basin with a 3 litre Chlorpyriphos solution (Chlorpyriphos 20 EC 7 ml/liter of water).
- To prevent the development of later stages of root grub, spray the arecanut palm with Imidacloprid 17.8 SL at a rate of 2.5 milligm per millilitre of water.
- The entomopathogenic worm *Steinernema carpocapsae* was applied in a liquid formulation at a rate of 1.5 billion infective juveniles per hectare.

White mite, *Oligonychus indicus*

Damage symptoms

- Nymphs and adults alike populate the underside of leaves, sucking the sap and leaving behind a bronzing pattern of yellow specks.
- The leaves turn yellow and eventually die if the infection is too bad.
- You can spot the colony on the underside of the leaves, just below the white web.
- In addition, the soft fruits are attacked, leading to deformities and dropping.
- Extreme conditions can cause seedling mortality.

Favorable conditions

- Typical of long periods of dry weather when activity levels are high.
- Mite infestation is common in palms because they thrive in environments with poor drainage and low organic and mineral materials.

Management

- Heavily contaminated, dried leaves can be removed and burned to eliminate the infection source.
- The mite population can be controlled by dusting with sulphur or spraying with 0.05% Wettable Sulfur.
- Apply a spray solution of 2 ml of Dicofol per litre of water (or 1.5 ml of Dimethoate) to the underside of infected leaves. If the bug problem returns after two weeks, spray again.
- Infestations of red and white mites can be controlled by spraying the undersides of leaves with a 0.025 percent solution of Trithion or Chlorbenzolate.
- Protect the predacious mites with the help of *Amblyseius channabasavanni*, the number of plant-eating mites can be kept in check.

DISEASES

Fruit rot (Kole roga), *Phytophthora arecae*

Symptoms

- The characteristic symptom is the widespread dropping of immature nuts, which thereafter decay and scatter at the tree's base.
- Early symptoms are water-soaked dark green or yellow lesions on the nut's surface close to the perianth (calyx).
- Before or after dropping their nuts, fruits can develop lesions that spread across their entire surface, leading to rot and a mycelial mass encasing the nuts as they decompose.
- As the disease progresses, it causes the fruit stalks and the axis of the inflorescence to dry up and rot, and it can even cause white mycelial mats to form on these areas.
- The term "Dry Mahali" refers to the rotting and drying up of nuts without shedding caused by an infection that occurs later in the season.
- Bud rot occurs when an infected heartleaf develops into a bud, while crown rot starts on the outermost leaf and works its way inside to the developing bud.
- Infected nuts lose their lustre and quality when they are green, reducing their market value.

Favorable conditions

- » Symptoms of the disease often begin 15 to 20 days after the start of typical monsoon rains, and they may last through the conclusion of the rainy season.
- » Low temperatures (about 20°C), high relative humidity (around 90%), and a lack of consistency in the amount of time spent in either rain or sun all contribute to the rapid spread of fruit rot.

Survival and spread

- » The pathogen can live in the form of latent mycelium or resting spores on the crown of the tree, on infected detritus, and on the soil's surface.
- » If diseased palms die and stay in the garden, they could serve as a primary inoculum source for the rest of the plants.
- » Diseases can be spread by high winds and rains by splashing them around.

Management

- » Pure plant growth (destruction of affected trees, collection and burning of fallen nuts).
- » To prevent damage to the grapes during the monsoon season, spray a Bordeaux mixture containing 1% sulfites over the clusters just before the rains start.
- » Prior to the onset of the monsoon's severe downpours, protect the areca bunches by covering them with polythene coverings (125-200 gauge, 24 x 30 inches).

Foot rot or anabe disease, *Ganoderma lucidum*

Symptoms

- » The outer whorls of the leaf first turn yellow, then the rest of the leaf, until the entire leaf is drooping and covering the stem.
- » Eventually, the leaves on the inner whorl will turn yellow, too. The leaves eventually wither and fall off, exposing the stem.
- » The stem then becomes weak and can be snapped by even moderate wind.
- » Brown discolouration and seeping of a dark fluid can be seen at the stem's base.
- » The bracket-shaped fruiting bodies of the fungus anabe appear at the tree's base.

- Dyscolored, dry, and brittle roots.
- Brown discoloration can be seen up to one metre from the ground when a diseased trunk is cut through.

Favorable conditions

- Disease flourishes on sandy and sandy loam soils, which become bogged down after heavy rains and dry out in the summer.

Survival and spread

- The pathogen survives in decomposing plant matter and spreads through the soil.
- The fungus travels from one plant to another by root contact or by landing on leftover garden trash. Water, excessive irrigation, and frequent ploughing and digging can all help disperse the infection.
- Secondly, the spores can disperse through the air.

Management

- It's important to replace any diseased or dead palms.
- Eliminate the source of the infection by removing the roots and stem.
- Trenches one metre deep and thirty centimetres wide can be dug to separate infected palms from healthy ones.
- Instead of flooding the fields with water in the summer, try using a basin irrigation system.
- Farmyard manure (50 kg) plus Trichoderma-enhanced Neem cake (5 kg) per palm every year.
- 20 litres of Bordeaux combination or Hexaconazole/Carbendazim diluted in water per palm tree, three times a year.
- Quarterly root feedings with either 2 g Aureofungin-Sol + 1 g CuSO4 in 100 ml of water, 2 ml Hexaconazole/100 ml, or 2 ml Calixin/100 ml. Only in the earliest stages of the disease will it have any effect.

Inflorescence dieback, *Colletotrichum gloeosporoides*

Symptoms

- The rachillae of the male flowers are the first to show signs of the disease, followed by brownish patches on the main rachis that quickly spread downwards to cover the entire rachis and ultimately cause wilting.

- Die-back symptoms are evident over the entire inflorescence, and the female flowers on the diseased rachis will fall off.
- Concentric rings in the discoloured portions represent the fungus' fruiting structures, called conidia.
- The loss of buttons and subsequent die-back of inflorescence is a major issue in monsoon-season arecanut plants.
- Palm death due to rot of terminal bud and surrounding tissues.

Favorable conditions

- The severity of the disease increases significantly in dry weather (February-March).

Survival and spread

- Conidia in the air cause the infection to spread.

Management

- Reducing the inoculum burden in the field is facilitated by removing and burning inflorescence.
- Spray with either Mancozeb (0.2%) or Copper oxychloride (0.25%).
- The 50 ppm concentration of aureofungin is similarly effective.

Bud rot, *Phytophthora palmivora*

Symptoms

- Spindly leaves turning from green to yellow and then brown is the first notice able symptom.
- Terminal bud and surrounding tissues decaying.
- Young whorls of leaves that are impacted can be removed easily by pulling.
- As the season progresses, the outer leaves turn yellow and fall off one by one, revealing the naked stem beneath.
- There's a bad smell coming from the rotting spindle.
- Palm dies with a severe infection.

Management

» To slow the progression of the disease and keep it from spreading, close planting, intensive intercropping, and low-lying farming should be avoided.

» Infected bud tissues are removed with a spoon and then treated with a 10% Bordeaux paste.

» Elimination and disposal of decaying palms and bundles.

» The spread of the disease could be reduced by soaking the palm tree crowns in a 1% Bordeaux mixture.

» Spraying a Mancozeb solution (5 g/300 ml of water) and placing two Mancozeb sachets (5 g/sachet) in the innermost leaf axils was proven to be an efficient prophylactic treatment for preventing disease.

» Phosphorus acid (Akomin) solution (0.5%) @ 300 ml/palm is a new environmentally friendly approach to the preventative control of bud rot disease.

» Created a plan for managing bud rot that includes harvesting diseased and dead palms, maintaining a clean growing environment, using fungicides and INM, preventing further outbreaks of pests and diseases, and more.

Yellow leaf disease, *Phytoplasma*

Symptoms

» First the inner whorl of leaves becomes yellow, then the rest of the crown.

» Finally, chlorosis is seen on nearly all leaves in the whorl, from the leaflets' margins to the midrib.

» The younger parts of the leaf first show signs of withering, and then the older ones catch up.

» The new leaves shorten and the laminae become uneven and floppy as they mature.

» Some instances of wilting and leaf drop have also been reported.

» Smaller and with blackened kernels, the nuts' production is drastically diminished.

» Infected palm trees develop a spongy, easily broken stem when the conducting strands are weakened.

» A top break occurs in a more advanced stage, when the stem is still alive.

- It's also been seen that the roots are rotting.
- The yield decreases over time.

Favorable conditions

- The occurrence of the disease is increased by the presence of water.

Survival and spread

- This virus is spread by *Proutista moesta*, a type of plant hopper.

Management

- To stop the spread of disease, dead plants must be removed.
- Improve the palm's health by applying 12 kilogm of FYM annually and providing summer irrigation and drainage.
- Planting tolerant types like True Mangala and South Kanara, which are resistant to blight.
- Excessive use of Potassium and Magnesium applications.

References

Daniel, M., & Kumar, T. P. (1976). Pests of arecanut. Journal of Plantation Crops, 4(2), 68-77.

Thube, S. H., Pandian, R. T. P., Bhavishya, E. K., Mohan, C., & Nagaraja, N. R. (2017). Major storage insect pests of arecanut Areca catechu L.-A Survey.

Mohan, C., Josephrajkumar, A., Thube, S. H., Saneera, E. K., & Rajkumar. (2022). Pests and Their Management in Arecanut. Trends in Horticultural Entomology, 1529-1544.

Nair, K. P., & Nair, K. P. (2021). Arecanut (Areca Catechu L.). Tree Crops: Harvesting Cash from the World's Important Cash Crops, 1-25.

Jose, C. T., Balasimha, D., & Kannan, C. (2008). Yield loss due to fruit rot (Mahali) disease of arecanut in Karnataka. Indian Journal of Arecanut, Spices and Medicinal Plants, 10(2), 45-51.

Harahap, D. (2022). Plant Diseases of Areca Nut. Intechopen

3 Betel Vine

INSECT PESTS

Aphid, *Aphis gossypii*

Damage symptoms

» Nymphs and adults alike desap the sensitive sprout and leaves of support plants like *Sesbania* spp., causing the leaves to turn yellow, curl, and wrinkly.

» Aphids produce honey dew, which lands on betel vine leaves and encourages the growth of sooty mould in the form of black patches.

Management

» Snip off any Sesbania leaves that have been infected with pests.

» Chlorpyriphos 2 ml/L should be sprayed on the agathi leaves.

» Aphids can be discouraged from feeding on plants by using reflective mulches, such as silver coloured plastic.

» Aphidids like *Aphelinus* and *Aphytis* and nematodes like *Diaeretiella rapae* are examples of parasitoids.

» The ladybird beetles *Coccinella septempunctata, Menochilus sexmaculatus, Hippodamia variegata,* and *Menochilus vicina* are among the species that prey on these insects. Syrphid fly, including the genera *Sphaerophoria, Eristallis, Metasyrphis, Xanthogramma,* and *Syrphus*. The Polish word for "lacewing" is "*Chrysoperla zastrowi sillemi.*" *Aphidoletes aphidimyza,* the common aphid midge. *Motacilla cospica* is a kind of predatory bird.

» Among the entomopathogenic fungi are the species *Cephalosporium, Entomophthora*, and *Verticillium lecanii.*

Scale insect, *Lepidosaphes cornutus*

Damage symptoms

- The leaves, petioles, and major veins are harbouring both nymphs and adults.
- The warty, crinkled, and eventually dried-out leaves that result from a scale infestation lose their original colour.
- The afflicted vines get sticky and eventually die off
- Yellowing of the leaves and wilting of the vine are the first signs.

Management

- You should only plant vines from seed that have not been infested with scale.
- Water containing 5% NSKE (50 g/L), 20% Chlorpyriphos (2 ml/L), or 50% Malathion (1 ml/L) can be sprayed.
- Neem oil diluted to 0.3 percent or Fish oil rosin soap diluted to 3.0 percent is also effective in preventing pest infestations in nurseries.
- Some examples of parasitoids are the citrine parasite *Encarsia* and the aphytid *Aphytis* sp.
- In the category of predators, we find Beetles such as the *Bdella* sp., *Aleurodothrips jasciatus, Karnyothrips melaleucus, Chilocorus circumspicus, C. nigrita, Pharoscymnus horni, Pseudoscymnus*, Lacewings, mirids, and Pytoseiids.

White fly, *Aleurocanthus nubilans, Dialeurodes pallida*

Damage symptoms

- Nymphs and adults both feed on the sap of young leaves, which can lead to yellowing, chlorotic patches, and sooty mould.
- The vines wither and the leaves turn chlorotic.

Pest identification

- The adult is a little bug with a white waxy bloom covering its body.

Management

- Use a 5% solution of neem seed kernel extract or a 2% solution of tobacco decoction in a spray bottle.
- *Cryptognatha* spp. (egg), *Encarsia* sp., *Eretmocerus* sp., and *Chrysocharis*

pentheus are all examples of parasitoids (nymphal).

» Mirid bugs (*Dicyphus hesperus*), ladybird beetles, lacewings, mites, ants, and the thrips species *Aleurodothrips fasciapennis* are among their natural enemies.

Mealy bug, *Ferrisia virgata*

Damage symptoms

» Insects can cause severe damage to young plants.

» Attack the plant's young and vulnerable parts, including as its leaves, stems, flowers, fruit, and roots.

» Nymphs and adults both feed on the sap from the leaves.

» Chlorotic leaves, failed flower buds, and little berries are the results of a severe infestation.

» Mold that looks like soot develops when honey dew is secreted (affects photosynthesis).

Pest identification

» The oval eggs are around 0.3 mm in length and have a shiny, pale yellow colour.

» The nymphs are covered in a white waxy substance and are yellow in colour. They have red eyes and an oval form.

» Its adult length is between 3 mm (females) to 4.5 mm (males). Females lack wings and range in hue from white to a pale brown; their legs and antennae are brown. Wax covers the entire body of an adult female, and there is a distinctive light grey stripe running down the back.

Management

» Use 2 ml/L of Chlorpyriphos 20 EC or Dimethoate as a spray. Spray the collar area where it is needed.

» Let use 10 insects per vine of the exotic predator *Cryptolaemus montrouzieri.*

Shoot bug, *Pachypeltis politus*

Damage symptoms

» Mirid bugs drain the moisture from young, fragile leaves and stems.

» As a result of being punctured and sucked dry, the leaves of plants are damaged by nymphs and adults.

- Both nymphs and adults feed on the sap of young, vulnerable leaves, causing the leaves to develop black spots. Adults are slim, energetic, and a reddish brown colour.
- The leaves' resale value drops as a result.

Favorable conditions

- Between June and September, the pest is most active.
- The pest flourishes in warm, humid environments.

Management

- Malathion sprays at 0.1% or pyrethrin sprays at 0.002% are both effective.
- After the leaves have been harvested, spray them with a 2% tobacco decoction, 0.5 % neem oil, 0.05% chlorpyriphos, or 0.05% monocrotophos solution.
- A 1.5% concentration of fish oil rosin soap is effective in controlling mirid bugs during severe infestations.
- Only 15 days after an insecticide has been applied can the leaves be collected. Before eating or selling the treated leaves, they must be thoroughly rinsed under running water.
- Effective pest control can also be achieved by spraying a solution containing 4% Neem seed kernel extract.

Red spider mite, *Tetranychus* sp.

Damage symptoms

- Spider mite nymphs and adults often use their long, needle-like mouth parts to eat the contents of the leaves' cells.
- Because of the decreased chlorophyll concentration, white or yellow spots will appear on the leaves.
- When an infestation is bad enough, the leaves wither and fall off.
- In severe situations, the mites also cause webbing on the leaf surfaces.
- Because of the high population densities, the mites migrate by forming a ball-like mass out of their silk threads, which is then carried by the wind to other leaves or plants (ballooning).
- The dry weather has helped the population increase.

Management

- Predators include Anthocorid bugs (*Orius* spp.), green lacewings (*Mallada basalis* and *Chrysoperla zastrowi sillemi*), predatory mites (*Amblyseius alstoniae, A. womersleyi, A. fallacies* and *Phytoseiulus persimilis*), predatory coccinellids (*Stethorus punctillum*), staphylinid beetle (*Oligota* spp.), predatory cecidomyiid fly (*Anthrocnodax occidentalis*), predatory gall midge (*Feltiella minuta*), etc.

DISEASES

Foot and leaf rot, *Phytophthora parasitica*

Symptoms

- Causes the leaves to wilt and turn brown, and the plant to collapse from foot rot.
- Dark brown, round spots occur at first, then spread and become rotten when exposed to high humidity for an extended period of time.
- In some cases, light brown zones form between dark brown necrotic patches.
- In the case of foot rot, the leaves quickly lose their sheen and the vines completely droop and dry up.
- The injured plant has suffered root damage.
- All of the roots' laterals have been cut off.
- The vines wilt and die off in clusters where the disease has taken hold.
- Over the course of a week, eighty to ninety percent of the vines will wilt and die.

Favorable conditions

- September through February, specifically, is optimal because of the high humidity and low nighttime temperatures (23 °C and below).
- Disease spreads more easily after July's rains.

Survival and spread

- The fungus is spread through the soil and lives on as a facultative saprophyte in decomposing infected plant matter and as oospores and chlamydospores in the soil.
- Irrigation water is the primary vector for the fungus's transmission from field to field.

» Splash irrigation and wind-borne rainfall facilitate the secondary dissemination of sporangia and zoospores.

Management

» An upgraded drainage system is a prerequisite for lowering disease severity.

» Diseased leaves and vines must be cut down and destroyed.

» Swap out your wheat for sorghum or corn every other year.

» A 1% Bordeaux mixture should be applied to cuttings before they are planted.

» Disease can be efficiently controlled by using a Bordeaux combination of 0.5-1.0% as a soil drench and a spray at 30- and 45-day intervals during the wet season.

» Copper oxychloride at a concentration of 4 gm per litre of water is an efficient method of disease control when applied to the soil.

» *Trichoderma harzianum* inoculated soil decreased vine mortality (11-21% versus 15-55% in control) and improved disease control.

Leaf spot/ anthracnose, *Colletotrichum piperis*

Symptoms

» Spots on leaves range in colour from light to dark brown and are encircled by a yellow, chlorotic halo.

» Black, necrotic tissue begins to spread from the leaf's periphery toward the leaf's core.

» On sometimes, a hazy yellow halo may appear.

» As anthracnose progresses, the vine's stem develops a ring of black, circular lesions that eventually kill the plant.

Favorable conditions

» Humidity and precipitation both play a role in the spread of disease.

Survival and spread

» The fungus survives in the field in the infected plant detritus.

» Conidia in the soil are the major vector of infection, and they can be dispersed by splashing water sources like rain or sprinklers.

» Conidia carried on the wind aid in the secondary spread in the field.

Management

- » Destruction of diseased leaves is an effective method for preventing further spread of the ailment.
- » Spraying with either a 1% Bordeaux mixture or 0.2% Zineb is beneficial in preventing the spread of anthracnose.

Sclerotial wilt / collar rot, *Sclerotium rolfsii*

Symptoms

- » Darkening of the plant's stem at its base, close to the earth.
- » Yellowing, sagging, and falling off of the leaves are the signs of leaf drop.
- » The entire vine eventually withers and dies.
- » Shrunken, squishy, and blackened, the affected areas of the stem shrank.
- » Fan-shaped, white ropy mycelial strands appear on the affected stem section.
- » Sclerotia, ranging in colour from brown to black, emerge on the diseased area.

Survival and spread

- » The fungus is spread through the soil and thrives on decomposing plant matter.
- » Likewise, the fungus can live as sclerotia in infected plant detritus in the soil for longer than a year.
- » Irrigation water carried the sclerotia from one area to another.

Management

- » Cut the infected vines down to the ground and burn the roots.
- » Neem cake, mustard cake, and farmyard manure are all excellent soil additions; use more of them.
- » To spray effectively, use either a 1% Bordeaux combination or a 0.2% Zineb solution.
- » Water the ground down with a Carbendazim solution of 0.1%.
- » Vine height increased from 67 to 69 cm to 199 to 254 cm when *T. harzianum* (strains MTCC Nos. 3841 to 3843) and *T. virens* (strain MTCC No. 794) were applied to the soil. Plant mortality fell from 73 to 87% to 7%, and disease control improved from 0% to 83%.
- » Significant increases in betel vine leaf yield were seen using the consortium of two strains of *Pseudomonas fluorescens* (NBRI-N6 and NBRI-N).

Powdery mildew, *Oidium piperis*

Symptoms

» Lower leaf surfaces have white to light brown powdery spots, which grow in size over time.

» To begin with, the infected leaves will have a few grey spots, which will get larger until the entire underside of the leaf is covered in a powdery mass of fungal growth.

» As the weather gets warmer, the leaf surfaces will become covered in a white floury mass of fungal development, eventually leading to premature leaf drop.

Favorable conditions

» Conditions ideal for the spread of disease are cool and gentle.

Survival and spread

» As dormant mycelium or haustoria encased in cell walls, the fungus can live on in the agricultural waste.

» Conidia carried by the wind cause secondary spread.

Management

» In order to prevent the disease from spreading, rotten leaves must be removed.

» Spray a solution of 2 gm of wettable sulphur per litre of water.

Bacterial leaf spot/ stem rot, *Xanthomonas campestris* pv. *betlicola*

Symptoms

» Black, irregularly shaped specks develop on the upper surface of the leaves, surrounded by a yellow halo.

» Canker of the vine stem manifests as elongated brown patches of varying sizes.

» When the infection is particularly bad, it causes leaf blight by cracking the stem and covering a considerable portion of the leaf lamina.

» Tiny water-filled lesions appear all over the leaf blade, their borders set by the veins.

» These eventually combine to form enormous, erratic brown patches.

» The leaves are prematurely falling off.

Favorable conditions

» High relative humidity and sporadic rain showers characterise the cloudy, damp weather.

» Vines are most vulnerable between the ages of two and three.

» The conditions for the spread of disease are ideal when temperatures and humidity levels are both high.

Survival and spread

» Viable bacteria in the diseased vines and leaves are the principal source of inoculum.

» The secondary spread is aided by rain splashes and splash irrigation water.

Management

» Replace the diseased vines and field weeds, and burn them.

» The disease can be efficiently managed with the application of either a 0.5% Bordeaux mixture or a Copper oxychloride solution (3 g/liter of water).

» After 20 days, spray with Streptocycline 400 ppm and the Bordeaux mixture at 0.25%.

References

Raut, S. K., & Bhattacharya, S. S. (1999). Pests and Diseases of Betelvine Piper Betle and Their Natural Enemies in India. Experimental & applied acarology, 23, 319-325.

Dasgupta, B., Roy, J. K., & Chitreshwar, S. (2000). Two major fungal diseases of betelvine. Diseases of plantation crops, spices, betelvine and mulberry, 133-137.

Balasubrahmanyam, V. R., Johri, J. K., Rawat, A. K. S., Tripathi, R. D., & Chaurasia, R. S. (1994). Betelvine (Piper betle L.). Betelvine (Piper betle L.).

Kumar, P., Das, S., Pandey, A. K., & Kumar, S. (2016). Major diseases of betelvine and their management. Crop Diseases and Their Management, 309.

Maiti, S., & Sen, C. (1982). Incidence of major diseases of betel vine in relation to weather. Indian phytopathology, 35(1), 14-17.

4 BLACK PEPPER

INSECT PESTS

Pollu beetle, *Longitarsus nigripennis*

Damage symptoms

» Infestations of the pollu beetle, found primarily in black pepper plantations in the plains and at lower altitudes, are common and destructive.

» The mature animals consume young plants, spikes, and berries.

» Blackening and eventual demise of infected shoots and spikes.

» After emerging, the grub bores into the berries, eating the insides until the berries are hollow.

» When pushed, the infected berries turn black and disintegrate.

» Shaded locations tend to have a higher pest density.

» Adults do not reproduce during the months of January through April, but they do stay in the field to feast on dead leaves.

Identification of pest

» Flea beetles, when fully grown, have a bluish yellow, metallic sheen.

Management

» Controlling how much sunlight reaches the plantation.

» In the months of June–July and September–October, you can spray either 0.05% Endosulfan or 0.05% Quinalphos.

» Apply a 0.6% solution of Neemgold (an insecticide derived from neem), or a 0.4% solution of Neem oil, during the months of August, September, and October.

» Management of shading and spraying with 0.6% Neem Gold every 21 days from June to October.

» The weaver ant, *Oecophylla smaragdina*, feeds on pollu beetles.

Top shoot borer, *Cydia hemidoxa*

Damage symptoms

» In newer plantations, the top shoot borer is more common.

» Moth caterpillars eat young, vulnerable shoots, causing them to shrivel and turn black.

» When young shoots on a growing vine are repeatedly damaged, the vine cannot expand.

» Infestation peaks between July and November, when the vines have the most tender new growth to offer to pests.

Identification of pest

» Eggs, despite their size, are colourless and transparent.

» 12-14 mm long, greyish green larvae

» The tiny, adult moth has black forewings with a crimson distal half and a greyish hind wing.

Management

» Applying a 0.05% solution of Monocrotophos or Endosulfan to the terminal buds of young plants and repeating the treatment once a month from July to October will keep the pest from damaging the plants' new growth as it emerges.

» Top shoot borer is preyed upon by the parasite insects *Apanteles cypris, Eudederus* sp. (Hymenoptera), *Goniozus* sp., and the parasitic mite *Clinotrombium* sp. (on larvae).

» The top shoot borer is a host for the entomopathogenic nematode *Hexamermis* sp.

Leaf gall thrips, *Liothrips karnyi*

Damage symptoms

» At higher elevations, leaf gall thrips infestation is more severe, especially on younger vines and in the nurseries where they are grown.

» Thrips cause marginal leaf galls by feeding on young, soft leaves, which causes the leaf margins to curl down and inward.

» Thicker, deformed, and crinkled, the infected leaves are the telltale sign of this pest.

» The insect gall contains evidence of the insect's development.

» A strong infestation can stunt the development of tender vines.

Identification of pest

» White, sluggish nymphs.

» A mature form characterised by wings that are extensively fringed.

Management

» Effective pest management can be achieved by spraying either Monocrotophos or Dimethoate at a rate of 0.05% when new flushes appear in the field or when cuttings emerge in the nursery.

» Parasitic mites including *Aphytis* sp., *Pseudoscymnus* sp., and *Chilocorus circumdatus* feed on leaf gall thrips.

» In the animal kingdom, predators include the likes of *Montandoniola moraguesi*, *Androthrips flavipes*, *Geogarypus* sp., *Lestodiplosis* sp., and *Rhodesiella* sp.

Scale insect, *Lepidosaphes piperis*

Damage symptoms

» The encrustations caused by scale insects are easy to spot on stems, leaves, and berries.

» Infested areas of vines will dry out and turn yellow because the insects are feeding on the plants' sap.

Identification of pest

» The scale is tiny, the colour is black, and the form is a boat.

» The adult males resemble small gnats and have only two wings and no mouth parts.

Management

- Remove and burn affected branches.
- Apply a 0.05% solution of Monocrotophos or Dimethoate at the first sign of an infestation and spray again 21 days later.
- Take care of shadows in the months of May and June, and in the months of October and November, spray Neem oil (0.6%).
- Mist a 2% Neem-Garlic suspension
- Two-week intervals of spraying with 0.05% Dimethoate.
- Several parasitic flies, including *Encarsia lounsburyi* and *Aphytis* sp., feed on the scale insect.
- Species of the mite *Bdella Aleurodothrips fasciatus, Karnyothrips melaleucus,* and Potential predators of scale insects include the ladybird beetle (*Coleoptera: Coccinellidae*), the lacewing (*Pseudoscymnus dwikalpa*), the earwig (*Hymenoptera: Sticholotisex anguis*), and the hornworm (*Hymenoptera: Pharoscymna shorni*).

Mealy bugs, *Pseudococcus virgatus* (*Ferrisia virgata*)

Damage symptoms

- Many mealybugs had settled in the vine's new growth and fruit.
- Fresh growth that is underdeveloped and stunted.
- Flawed produce.
- Leaves are turning yellow.
- Honeydew is a sticky material that insects exude; it is responsible for encouraging the secondary growth of sooty moulds.
- Sap sucking causes the plant to dry out and lose its leaves and branches, and ultimately turn a yellow colour.
- Many root mealy bug-infested vines will also have Phytophthora and nematode problems.

Identification of pest

- Eggs fluctuate in colour from orange to a yellowish hue.
- Soft-bodied insects are slow to move and covered with waxy white threads on the females.
- Mealy bug female adults are a rosy white colour and are only lightly waxed.

In their early stages, male and female mealy bugs look very similar to one another.

Management

- Spraying the plants with a vigorous jet of water may expel the insects.
- If the infestation is really bad, a spray of 0.05 percent Monocrotophos will do the trick.
- Soil drenched with 0.075% Chlorpyriphos is effective against mealy insect infestations on roots.
- Mealy bugs are the prey of the coccinellid, *Cryptolaemus montrouzieri.*
- Ants are a common source of care for mealy bugs.

DISEASES

Quick wilt/ Foot rot *Phytopthora capsici*

Symptoms

- Leaves may develop water-soaked lesions on the borders or in the middle, and these lesions will have a fimbriate margin.
- Defoliation happens when a disease is particularly bad.
- Infection can spread to aerial limbs at any node.
- Discoloration and rotting spread upwards and downwards from the point of infection on the branch, causing die-back.
- The damaged vines' lateral shoots die off at the nodes.
- Within two to three weeks, the infected portion of the vine's stem near the ground rots and dies.
- Odor is being released from the affected area.
- The rotting tissue travels down the stem and into the roots, where it causes further problems.
- Yellowing and dropping of the leaves is a symptom.
- Both the main root and the feeder root might get infected.

Management

- Vines that are seriously infected should be uprooted from the ground and destroyed.

- Grow cuttings in biocontrol agents like *Trichoderma* spp. and *endomycorrhizae*-enhanced solarized or steam-sterilized soil.
- Maintain proper drainage.
- Pruning the runner shoots and leaves at ground level protects them from the harsh environment by keeping them off the ground.
- It is recommended to apply *Trichoderma* spp. at 75 g/vine (containing 108 cfu/g of formulation) in combination with coffee husk and well-rotten cow dung in the months of May and June (pre-monsoon in the months of August and September), followed by a second treatment of Neem cake at 2 kg/vine (post-monsoon).
- In the months of May and June (just before the monsoon), you should rub Bordeaux paste into your neck and chest, and then spray either 1% Bordeaux combination, 0.2% Copper oxychloride, 0.01% Metalaxyl, or 0.3% Potassium phosphonate.
- With the commencement of pre-monsoon showers in May-June, irrigate plant basins with Copper oxychloride 0.2% @ 5-10 L/vine and spray Bordeaux mixture 1% on the new foliage; this process should be repeated in August-September and October.
- Apply *Pseudomonas fluorescens* (50 g/vine, 108 cfu/g of formulation) and *Trichoderma harzianum* (IISR 1369) together (IISR-41).
- Kaluvally was the variety with the lowest disease incidence (5.1%).

Fusarium yellows, *Fusarium solani*, *F. proliferatum*

Symptoms

- Roots rot, which weakens plants and turns their leaves yellow.
- Root rotting.
- It doesn't take long for the damaged plants to perish.
- It has been found that plantations infected with fusarium have a shorter economic life span (from 20 years to 6-8 years) and produce less fruit (from 3 to 1.5 kilogm per plant).

Management

- To prevent the spread of disease, farmers use phytosanitary methods such as the destruction of diseased plants through burning and the application of 200 gm of calcium cyanamide per mound.

- Benomyl or Thiobendazole 500 ppm sprayed in the nursery.
- The soil and plant material should be treated with Benomyl (0.05%) and Thiobendazole (0.06%) every 14 days as a preventative measure against fungal diseases.
- The *Trichoderma harzianum* CH-1 strain was found to be efficient against Fusarium wilt (controlling the disease by 78.3%).

Anthracnose, *Colletotrichum gloeosporoides*

Symptoms

- Chlorotic leaf spots are angular in shape and have a yellow halo around them.
- The infected spikes wither, become black, and fall off.
- Infected berries progressively dry out and continue to have a light appearance.
- Necrosis on the spike stalk brought on by the fungus causes the spike to fall off.

Management

- Watering grapevines in the heat of summer.
- Controlling the amount of available shadow is essential.
- Spraying with a 1% Bordeaux combination before and after a monsoon is recommended.
- Sprays of 4.5 g/L of prochloraz manganese chloride (50%) are applied three times yearly, at 14-day intervals, when there is 50% berry spike emergence.
- With a maximum of 6 applications each season and a 14-day interval before harvest, carbendazim (50%) is sprayed at 0.65 g/L.

Stunted disease, *Cucumber Mosaic Virus* and *Badna Virus*

Symptoms

- Observed in the Indian states of Kerala's Kodagu, Hassan, and Uttara Kannada and Karnataka's Kodagu and Idukki districts.
- Internode shortening and plant stunting are observed in the vines to varied degrees.
- Deformation causes the leaves to shrink and narrow; they also take on a leathery, puckered, crinkled appearance.

- On sometimes, chlorotic patches and streaks will also emerge on the leaves.
- Light and dark green regions, or green and yellow, appear mosaic-like on the affected vine.
- The vines are stunted because of the disease.
- When a vine loses all of its leaves and fruit, it is called a "berry vine."
- Infected vines gradually lose production.

Spread

- Viral propagation primarily occurs via the transplantation of infected stem cuttings.
- Aphids and mealybugs are examples of insects capable of spreading the disease.

Management

- Infected plants, especially at the row's ends, should be rogued out before the infection may spread further.
- Key weeds should be eliminated before planting crops.
- Plants susceptible to the virus have been protected through intercropping with maize or other tall, resistant barrier crops.

References

Krishnamoorthy, B., & Parthasarathy, V. A. (2010). Improvement of black pepper. CABI Reviews, (2010), 1-12.

Adama, A., Ee, K. P., Sahari, N., Tida, A., Shang, C. Y., Tawie, K. M., ... & Mohamad, H. (2018). Dr. LADA: Diagnosing black pepper pest and diseases with decision tree. International Journal on Advanced Science, Engineering and Information Technology, 8(4-2), 1584-1590.

Keuh, T. K. (1990). Major diseases of black pepper and their management. Planter, Kuala Lumpur, 66(767), 59-69.

Babu, B. S., Pandravada, S. R., Rao, R. P., Anitha, K., Chakrabarty, S. K., & Varaprasad, K. S. (2011). Global sources of pepper genetic resources against arthropods, nematodes and pathogens. Crop Protection, 30(4), 389-400.

Kenyon, L., Kumar, S., Tsai, W. S., & Hughes, J. D. A. (2014). Virus diseases of peppers (Capsicum spp.) and their control. In Advances in virus research (Vol. 90, pp. 297-354). Academic Press.

5 Cardamom

INSECT PESTS

Thrips, *Sciothrips cardamomi*

Symptoms

- Nymphs and adults use their rasping-sucking mouth parts to consume tender leaf sheaths and the tips of unopened flower bracts.
- Silvering of the injured area.
- Panic attacks weaken over time.
- Flowers and immature capsules fall off, causing a drop in the amount of mature capsules.
- Corky encrustations develop on the capsule when they are infested, leading to the capsule's deformity and shrivelling.
- These pods don't have that wonderful fragrance, and the seeds within aren't fully matured.
- The contaminated pills are cheap because they are of poor quality and have a low weight.

Identification of pest

- Eggs are placed singly in the soft section of the leaf sheath and have a kidney shape.
- The nymphs are quite little, delicate, and straw-yellow in hue.
- Adults measure only 1.25-1.50 mm in length and have fringed wings of a dark greyish brown colour.

Management

- » Controlling the amount of shadow present in the area and removing potential hosts from the environment are two such measures.
- » Elimination of the dried shoot skins at the stem's base.
- » Between March and September, you can choose from four sprayings with either 0.025% Monocrotophos, 0.05% Dimethoate, or 0.025% Quinalphos.
- » Use of rosin soap containing fish oil or permethrin.
- » Mudigere-1 (Malabar) cardamom is resistant to thrips.
- » Both *Orius laevigatus* and *Thripoctenus americensi* feed on thrips as a primary food source.

Shoot/panicle/capsule borer, *Dichocrocis punctiferalis*

Symptoms

- » This yellow moth's caterpillars kill the crop's centre spindle early on, when the crop is still young, and the outcome looks like a bunch of dead hearts hanging from the aerial stem.
- » In order to get at the seeds inside, caterpillars will bore through the capsules.
- » Caterpillar presence in the pseudo stem, inflorescence, and pods is indicated by the presence of excreta at the area of assault.

Identification of pest

- » Pink, oval, and flat eggs are placed individually or in small clusters on the young plant tissue.
- » The larva is long, pale green and pink on the back, with a brown head and pro-thoracic shield and fine hairs all over its body.
- » The adult moth is a manageable 22–24 millimetres in size, and its wings are a pale golden brown with black markings.

Management

- » Adult moths can be found dozing on the underside of a cardamom pod throughout the day. It is possible to catch them with an insect net and destroy them.
- » Care must be taken while cutting down tillers that are exhibiting 'dead heart' symptoms.

- If the attack is recent, as seen by the lack of frass extrusion, the tillers can be cut.
- Use a spray containing either 0.025% quinolone, 0.1% carbaryl, 0.05% monocrotophos, 0.05% fenthion, 0.05% dimethoate, 0.1% endosulfan, or 0.1% phenthoate between February and March and September and October.
- To inject a *Bacillus thuringiensis* preparation into a bore hole (0.5 ml in 10 ml water).
- Many different parasitic insects, such as *Trichogramma chilonis, Tetrastichus* spp.,*Apanteles* spp., *Eriborus* spp., *Friona* spp., *Telenomus* spp., *Xanthopimpla australicus*, and *Agrypon* spp., feed on the pest.
- Mudigere-1(Malabar) cardamom is resistant to shoot borer.

Rhizome weevil, *Prodioctes haematicus*

Damage

- Between the months of November and January, this is a major problem in the secondary nursery, especially in areas where seedlings are routinely grown year after year.
- Cardamom clumps as a whole will die if grubs are allowed to tunnel and feed on the rhizome and the lower part of the stem.
- The leaves wither and the stems wither and snap.

Identification of pest

- The elytra of an adult have three black dots, two in front and one in behind, while the pronotum has three black lines, one in the middle of the dorsum and the other two on either side.

Management

- Replace the plant or seedlings if they are affected.
- In order to replace the bug, Chlorpyriphos 0.04% drenched into the soil of the nursery beds should do the trick.
- Apply a solution of Malathion 1.25 L or Carbaryl 50 WP 1.25 kg in 500-1000 L of water/ha directly to the base of the cluster.

Aphids, *Pentalonia nigronervosa*

Damage

- Aphids spread the Katte virus.
- Adults and nymphs alike populate the false stem and leaf sheath.
- Aphid colonies live in the leaf sheath of decaying pseudo stems.
- Nits and adults alike drain a plant's vitality by sucking its sap.

Identification of pest

- Darkness characterises nymphs.
- The wings of adult birds are brown with black veining.

Management

- Bananas should not be planted anywhere near or inside cardamom plantations.
- Remove any additional potential canna, alocasia, colocasia, etc. host plants from the area.
- It's important to replce the aphid colonies, therefore cut off any pseudo stems that have dried out and rotted.
- Applying a spray of 0.05% Methyl Demeton or Dimethoate is an efficient method of controlling aphids.
- *Aphidius* and *Aphelinus* species are parasitoids that feed on aphids.

Root grub, *Basilepta fulvicorne*

Damage symptoms

- Grubs' uneven scraping of cardamom roots is a sort of feeding.
- The first sign is leaf yellowing, followed by drying, rotting, and ultimately death of the plant.
- The leaves of badly affected plants become dry and yellow.

Identification of pest

- Cardamom tracts are especially vulnerable to injury from *B. fulvicorne.*
- The typical hue of an egg is a light yellow.
- Grubs can be identified by their short, stocky, 'C' form and off-white coloration.

» Beetles can be any of several metallic hues, including blue, green, greenish brown, and brown.

Management

» During the major emergence periods (March–April and August–September), collect the beetle with hand nets or sticky traps and eliminate.

» Drenching with Chlorpyriphos 0.04% at 3-4 litres per clump or applying Phorate at 2-4 g a.i. /clump 10-15 cm surrounding the plant is effective for controlling the grub's early stages, which are typically present in soil between May and June and September and October. Successful eradication of root grubs requires a light raking of soil prior to the application of insecticide.

» When Carbofuran 3G (at 100 g/plant) and Imidacloprid (at 0.5 ml/liter of water) were applied to the soil in a cardamom field, the number of root rot grubs dropped dramatically.

White fly, *Kanakarajiella (Dialeurodes) cardamomi*

Damage

» They are a major problem in the areas of Kerala that are used for growing cardamom.

» Nymphs and adults feed on the undersides of leaves, where they cause yellowing, staining, and drying by sucking the sap from the leaves.

» Honey dew covers the stunted, infected plant.

» As the black sooty mould spreads, it prevents the leaves from producing oxygen and so stops photosynthesis.

Identification of pest

» The cylindrical eggs have a creamy yellow colour when first laid and develop a brown hue with time.

» Nymphs are round and very light green. The nymphal life cycle consists of four distinct forms.

» The adults have two sets of white wings and a petite, delicate body, measuring only around 2 mm in length.

Favorable conditions

» Pest populations tend to increase during the warmer months.

Management

- Nymphs and puparia should be gathered and used to eliminate damaged leaves.
- Set up 12 yellow sticky traps per hectare.
- Spraying the underside of the leaves with a solution of 500 ml of Neem oil and 500 ml of Triton in 100 litres of water is an excellent way to replace nymphs.
- Methyl Demeton 25 EC, Dimethoate 30 EC, 1.0 litres of water, or 500 gm of Acephate 75 SP per hectare should be sprayed on the land. Two or three applications of the spray every 15 days is possible.
- Whiteflies are preyed upon by parasites such as the *Encarsia*, *Eretmocerus*, and *Chrysocharis* species.
- *Malleada boninensis*, more commonly known as lacewings, feeds primarily on whiteflies.

DISEASES

Rhizome rot / Clump rot, *Pythium vexans*

Symptoms

- Eventually, the young leaves of infected plants will turn yellow and wither and die.
- The rhizomes rot at the stem's base, causing premature death of older leaves and the weak, decaying shoots that emerge in their place.
- It's simple to remove the infected shoot.
- The presence of dead or dying shoots on the stem is a breeding ground for the disease.

Management

- Elimination of affected clusters.
- Suitable drainage is provided.
- Using *Trichoderma harzianum* in soil and solarizing nursery beds.
- Incorporating solarization of the nursery mixture and the use of AMF/*T. harzianum* (cfu/g) while packing nursery soil into plastic bags.
- An application of 3 litres of Bordeaux mixture per plant, at a rate of 1%.

- The combination of *T. harzianum* (IISR 1369) and *Pseudomonas fluorescens* (IISR- 11 or 6).
- Tolerant strains include ICRI-2 (Mysore) and ICRI-3 (Malabar).

Capsule rot/ Azhukal disease, *Phytophthora parasitica* var. *nicotianae*

Losses

- Cardamom production in Kerala's Idukki district drops by 30 percent due to this disease.

Symptoms

- Water-soaked black dots emerge on leaves in large, uneven circles.
- A section of the unopened leaves that was left exposed has begun to decay.
- The underside of the leaf sheath develops a grey pattern of irregular dots bordered by brown.
- In this case, the base rots and the pseudo stem separates at the collar.
- The rhizomes decay as the infection travels downward into the plant.
- Tender green fruits develop tiny light brown blemishes and fall off after three to five days, leaving only the tiny fruit stalk.
- The inflorescence rots at the very tip.

Favorable conditions

- The low temperatures and excessive humidity found in the plantations are ideal for the spread of disease.

Survival and spread

- The chlamydospore is the infectious form of the bacterium.
- For 32 weeks in air-dried soil and 42 weeks in moist soil, the fungus can live in infected rhizomes.
- The majority of an infection's inoculum comes from contaminated soil and plant matter.

Management

- Disposal and destruction by fire of diseased tissues and roots.
- Don't take rhizomes from infected areas and plant them in healthy soil elsewhere.

- Repair the drains.
- Before the rains start, spray the young plants with a 1% Bordeaux mixture containing sticker. The use of sprays might be extended until November/December.
- Spray the panicle heavily (3 litres of Bordeaux mixture per plant in July and August) to promote blooming.
- It was also discovered that spraying with 0.25 percent Metalaxyl MZ or 1 percent Fosetyl AL was effective.
- *T. viride, T. harzianum*, and *Bacillus subtilis* were introduced into the soil.
- Mysore and Malabar cultivars have fair resistance.

Leaf blight, *Colletotrichum gloeosporioides*

Symptoms

- The upper surface of the leaf develops elongated, water-soaked lesions of varied sizes.
- The dots darken to brown with a faint yellow centre.
- The leaves fall off and the pseudo stem dies.
- Any new growth that emerges is stunted and fails to fully mature.
- Beginning at its tip, the inflorescence dries up.
- The burned aspect of the gardens is clear.

Favorable conditions

- Misty and rainy weather that comes and goes.

Survival and spread

- In open places with little to no shade, the disease is reported to be more virulent.
- The symptoms, which first occurs around the monsoon's midpoint, reaches its peak in the late monsoon.

Management

- The contaminated leaves must be removed and disposed of.
- There should be enough shade for everyone.
- Carbendazim, Mancozeb, or Copper oxychloride sprayed at 3 g/L in water

once a month.

Katte/Marble disease, *Cardamom Mosaic Virus* (Potyvirus)

Losses

» Pure cardamom crops experienced 38%, 62%, and 68% crop losses in the first, second, and third years following infection.

Symptoms

» Chlorosis is a common condition in young leaves, manifesting as narrow, pale green bands that run parallel to the veins from the leaf's midrib to its margins.

» There are also stitching stripes on the leaf sheath.

» When the disease has progressed, it causes the entire plant to exhibit mosaic symptoms, clumps to quickly disintegrate, and plants to be abnormally short.

» The plant's rhizome withers and dies.

» Attaching young clumps results in their demise before they can bloom.

Transmission

» The *Pentalonia nigronervosa* banana aphid is responsible for spreading this virus.

Spread

» Infected rhizomes are the primary vector for the virus's transmission.

Management

» Collecting and burning contaminated clumps, together with any accompanying rhizomes, is a common method of eradication.

» Crops for the nursery are grown in disease-free regions.

» Insecticide spraying (with 0.05% Dimethoate or Phosphamidon) to eliminate the vector.

» One of the most effective predators of the aphid is the green lacewing, or *Chrysoperla carnea*.

Nilgiris necrosis, *Nilgiris Necrosis Virus*

Symptoms

- Leaves turn a reddish brown colour after first appearing with green or yellowish streaks that run parallel to the veins from the leaf's midrib to its margins.
- Disjointed spots also manifest on different leaf surfaces.
- The tillers are all infected because the pathogen is systemic.
- The plants' growth is limited and their leaves shrink.
- The adult leaves have a torn appearance and rip off along the necrotic patches.
- The disease causes stunting and an inability to produce panicles and capsules in affected plants.

Survival and spread

- Infected seeds and potted plants are the primary vectors for disease transmission.

Management

- Plant healthy rhizomes.

References

Yadav, P. K., Basnet, B., & Devkota, A. R. (2021). Insect pests infestation, diseases and management practice of large cardamom in Nepal: A Review. INWASCON Technology Magazine (i-TECH MAG), 3, 09-18.

Ravindran, P. N., & Madhusoodanan, K. J. (Eds.). (2002). Cardamom: the genus Elettaria. CRC Press.

Rajan, P. (1981). Biology of Pentalonia nigronervosa f. caladii van der Goot, vector of'katte'disease of cardamom. Journal of Plantation Crops, 9(1), 34-41.

6 Cashew

INSECT PESTS

Stem and root borer, *Placaederus ferrugineus*

Damage symptoms

- » The delicate tissues of bark are consumed by the borer rub.
- » Small bore holes can be seen in the collar area.
- » Frass (like coarse dust powder) and excreta are expelled from the collar region holes, and the bark becomes discoloured as a result.
- » Gummosis, or the seepage of gum, at the cashew tree's trunk base.
- » When a tree is damaged, it stops receiving its normal supply of water and nutrients, causing its leaves to turn yellow and fall off.
- » Loss of anchoring causes the trees to lean to one side.
- » Infestations typically occur on tree trunks up to 100 centimetres off the ground.
- » Yellowing of the leaves and subsequent drying of the limbs and eventual death of the tree stem from a lack of preventative action.

Pest identification

- » Eggs are roughly 3 mm in length, white, and ovoid (looks like rice grain).
- » The adult grub has an off-white tint.
- » The adult beetle has reddish brown long horns and measures between 25 and 40 mm in length. The male antenna is about twice as long as the female's.

Management

- Carbaryl 0.2% @ 4 g in 1 litre water should be used to swab the trunk area up to 1 m from the ground as a preventative measure.
- At least once every six months, replace any trees that are dead or infested.
- Apply Chlorpyriphos (10 ml in 1 L water) to the infected trees' trunks and exposed roots up to 1 m from the ground.
- Closing the flap after stuffing the stem with cotton-wool soaked with Monocrotophos (30 ml).
- Chlorpyriphos @10 ml in one litre of water or Carbaryl 0.2% solution @4 g in one litre of water should be drenched into the soil surrounding the tree trunk.

Tea mosquito bug, *Helopeltis antonnii*

Damage symptoms

- Both juveniles and adults feed on the sap of newly developed flowers, leaves, fruits, and nuts.
- Dark brown spots appear on the green, fragile stems of new shoots and the rachises of inflorescences, and in some cases, gum is secreted as well.
- Crinkling and curling are the results of feeding on young, fragile leaves.
- Trees with a lot of infestations seem burned and lose their developing tips and sprouts.
- Dieback occurs in severely injured stems.
- As a result, several tertiary buds form, leading to the characteristic "witches' broom" appearance.

Pest identification

- Reniform, white eggs have a smooth texture.
- Nymphs tend to be a fiery orange colour.
- Approximately 6-8 mm in length, the adult reddish-brown beetle has a black head, red thorax, and black and white abdomen.

Management

- Cashews should not be planted next to cotton, tea, or sweet potatoes since these plants serve as hosts for the *Helopeltis* bug.

» Spray as directed: at the beginning of the new flushing stage, spray with either 1.5 ml/liter of water (0.05%) Monocrotophos or 0.003% Lamda cyhalothrin (November- December). At the flowering stage, spray again with Carbaryl at 2 g/l (0.1%) or Chlorpyriphos (0.05%). (Dec- January). At the beginning of the fruiting process, the third spray is a duplicate of the first (Feb-March).

» Cashew trees are well-protected by weaver ants because they construct nests there.

» Both the *Crematogaster wroughtonii* and the *Oecophylla smaragdina* are known predators.

» *Erythmelus helopeltidis, Telenomus cuspis, Chaetostricha* sp, and *Gonatocerus* sp. nr. *bialbifuniculatus* are all parasitoids.

» Pest infestation was rather low in commonly grown cashew cultivars like Dhana, Bhaskara, and the accession VTH 153/1.

Flower thrips, *Scirtothrips dorsalis*

Damage symptoms

» Both adults and juveniles feed on the sap of leaves, flowers, and fruit and vegetable crops by clustering on the undersides of leaves.

» Their feeding causes the foliage to turn a drab brown and causes scab on flower stems, fruit trees, and nuts to develop a corky layer.

» The terminal leaves begin to curl inward from the margin toward the mid-rib after being continuously fed on by a high number of thrips.

» Trees will drop their leaves and slow down their growth if the condition is severe.

» Nuts that aren't fully ripe may drop their hulls or have an uneven fill in the kernel if they're attacked by thrips.

» Nuts with the disease are unsafe to eat.

Pest identification

» The eggs are hyaline and reniform (kidney-shaped) when they are laid, but they turn a pale yellow right before they hatch.

» Nymphs lack wings and are a pale yellow colour.

» Adults are quite tiny, measuring just around 1 mm in length, and have a dull yellowish brown colour.

Favorable conditions

- Reduced leaf wetness and severe drought are both conditions that promote rapid thrips growth.
- Hotter temperatures during the right time of year allow its population to expand.

Management

- Thrips can be effectively controlled by spraying a solution containing 0.03% Lambda Cyhalothrin, 0.03% Phosphamidon, 0.03% Dimethoate, or 0.003% Methyl-Demeton.
- It was also shown that spraying flowers with Monocrotophos 36 WSC 0.05% (@ 1.5 ml/liter) was efficient in controlling flower thrips.

Apple and nut borer, *Thylocoptila panrosema*

Damage symptoms

- The immature larvae first go to the joints of the nut and apple, where they scrape the skin and bore into the fruit.
- In their later stages, they transform into borers and feed on soft fruits and nuts.
- Caterpillars eat the fruit at every stage: immature apples and nuts fall from the trees because the nuts never fully develop.
- The borer has dug a tunnel close to where the apples and nuts meet, and its exit points are blocked with faeces.
- However, there have been tales of as many as five caterpillars in an apple and three in a nut.
- The caterpillar ate its way through the apple.

Pest identification

- The adult moth is about the size of a human thumb and has pale, black wings.

Management

- In the pre-flowering season, it is effective to completely remove and destroy all dried inflorescence as a cultural approach of pest control.
- At the period of fruit setting, spray Carbaryl 50 WP 0.1% (@ 2 g/liter).
- Lambda Cyhalothrin at 0.6 ml/liter or Dichlorvos at 2 ml/liter of water spraying during the off season.

Mealy bug, *Ferrisia virgata*

Damage symptoms

» Mealy bug nymphs and adults feed on the sap of young plants, stunting the development of shoots, flowers, and fruits.

» It manifests on the undersides of young leaves, twigs, panicles of flowers, and the stalks of fruits.

» Beetles not only harm plants directly, but their honey dew attracts sooty mould, which stunts photosynthesis.

» In the midst of a major epidemic, nut yields dropped drastically.

Pest identification

» The yolks of eggs have a warm, golden hue.

» Flat, very little, oval, and yellow nymphs are covered in fluffy, white wax.

» An adult has a milky white coating all over its body and has a soft, pliable build.

Management

» Manage this pest by keeping a close eye out for signs of infestation and catching them early.

» Weeds and other potential pests should not be growing in or around the cashew plantation.

» Mealy bug colonies can be removed from plants by removing and discarding infested portions, such as leaves, inflorescence, and twigs.

» If you want to stop the spread of a pest, you should gather up the leaves that have fallen under a tree's canopy and burn them.

» Using a spray containing either 0.2% Dichlorvas 76 WSC (at 2.5 ml/liter), 0.05% Methyl Parathion (at 1 ml/liter), or 0.05% Dimethoate (at 1.75 ml/liter) in combination with 20 g of fish oil resin soap per litre of water has been shown to significantly reduce the number of insects present.

» The predators, such as the *Chrysoperla carnea, Menochilus septumpunctata*, and *Scymnus coccivorus*, should be used.

Leaf miner, *Acrocercops syngramma*

Damage symptoms

- Once the eggs hatch, the larvae begin feeding on the epidermal layer of the young cashew leaves.
- Where multiple mined sections have merged into one another, the epidermal cuticle will expand and cause the leaf surface to seem white and blistered.
- Because of the insects' eating habits, the affected areas develop greyish white blistering patches.
- Leaves with pest infestations show severe damage in the form of gaping holes as the leaves grow.
- Newer plantations and seedlings in nurseries are more likely to be infested by this insect.
- The mined areas of the leaf's surface will dry out and crumble.

Pest identification

- Larvae are colourless when first hatching and transform into a reddish brown caterpillar as they mature.
- The adult moths are a metallic grey colour.

Management

- It is suggested that infested cashew leaves be removed and destroyed to stop the spread of the pests.
- At the new flush emerging stage, which occurs in the months of October and November, spray Monocrotophos 36 WSC 0.05% (@ 1.5 ml/liter).

Leaf and blossom webber, *Lamida moncusalis*

Damage symptoms

- New flushes and panicles bring with them the phenomenon of leaf and flower webber.
- The first sign of an infestation is the appearance of webs on the tips of young stems and blossoms.
- Coevolving web shoot and inflorescence in caterpillars.
- They stay within the web and consume it.

- Webs are also used to coat and scrape the surfaces of apples and nuts.
- The nut fails to germinate and cracks.
- Any bullet fired into the webbed section quickly dries up.
- Most severely affected are young trees.

Pest identification

- The colour of newly deposited eggs is a pale golden green.
- The caterpillar is a dark green colour with golden longitudinal bands and rosy dorsal lines.
- The adult moths are a metallic grey colour.

Management

- The contaminated leaves must be removed and disposed of.
- The cocoons can be destroyed or buried by digging around the base of the cashew tree.
- Effective pest control can be achieved by spraying a solution of 2 ml of pesticide per litre of water containing monocrotophos, profenophos, lambda cyhalothrin, endosulfan, or chlorpyriphos.
- Braconidae and Elasmidae, two families of parasitoid insects, were determined to be responsible for between 10 and 50 percent of all parasitism.

Weevil, *Mecocorynus loripes*

Damage symptoms

- The larvae eat the sapwood and tunnel downwards through the bark.
- The main trunk and major branches are covered in a brownish-black sticky frass.
- Trees that are heavily attacked suffer sapwood damage along their circumference and eventually die.

Pest identification

- Larvae are wingless grubs that are mostly white and have a brown head.
- The cashew weevil is an enormous weevil, measuring 20 mm in length and sporting a distinctive knobby look. The colour is a murky grey-brown.

Management

- Lightly affected trees can have their bark stripped to expose the larvae and pupae underneath, which can then be killed. If necessary, continue doing this once a month for an additional six months.
- Kill trees that have become seriously infected. First, you must find and kill all of the adult weevils. Then, you must cut down the tree, peel back the bark to reveal the galleries where the larvae are hiding, and last, you must kill the larvae and the pupae before you can burn the tree.

Red-banded thrips, *Selenothrips rubrocinctus*

Damage symptoms

- Both adults and juveniles feed on the sap of leaves, flowers, and fruit and vegetable crops by clustering on the undersides of leaves.
- Their feeding causes the foliage to turn a drab brown and causes scab on flower stems, fruit trees, and nuts to develop a corky layer.
- In extreme circumstances, trees may lose leaves and experience growth stunting.

Pest identification

- Wings of mature individuals are fringed.
- The first and last abdominal segments of nymphs are red-banded, giving them a greenish yellow overall appearance.

Management

- Spreading Monocrotophos 36 WSC 0.05% at 1.5 ml/L in a spray bottle.
- Using a dilution of 1 millilitre of active ingredient per litre of water to spray either dimethoate 30 EC or methyl demeton 25 EC.

DISEASES

Damping-off, *Pythium* sp., *Rhizoctonia* sp.

Symptoms

- When the soil medium in a nursery becomes excessively damp, it has a negative impact on cashew seedlings.
- The seedling is infected before it emerges from the soil, a process known

as pre- emergence damping off. Pre-emergence damping off diseases mostly target seed nuts because of the large amounts of amino acids and carbohydrates they emit during germination.

» Post-emergence damping off occurs when young seedlings are infected in the collar area.

» Seedlings suffer catastrophic losses during the nursery stage when they are attacked by fungi, either in the collar region or the root zone, causing the stem to turn pale and discoloured from being saturated in water.

Favorable conditions

» In a nursery with poor drainage, the disease incidence is high.

» Dampening disease is facilitated by close spacing, continuous nursery growing, chilly, wet, highly organic, neutral to basic soils.

Management

» In nurseries, this means ensuring adequate watering and drainage.

» It is recommended to use Arasan on planting seeds before planting them.

» When potting, it's important to use soil that has been treated with soil fumigants.

» Drainage needs to be properly addressed before the rainy season begins.

» You can stop the spread of the disease by dousing the bags in a 1% Bordeaux combination, 3 g/L of Copper oxychloride, or 3 g/L of Mancozeb solution.

Anthracnose, *Glomerella cingulata* (*Collectorichum gloeosporoides*)

Symptoms

» The fungus causes watery, reddish brown lesions on the young, susceptible parts of the plant, such as the leaves, twigs, inflorescences, fruits (apples), and nuts.

» You can see resin oozing out of the damaged area.

» As the sores grow, the young plants die.

» Normally juicy fruits and nuts have shrunken, and the sensitive leaves have crinkled.

» Infected flower buds become dark in colour.

» As the disease progresses, the leaves and little fruits eventually fall off.

» If the tree's terminal shoots are repeatedly infected, the tree will eventually die.

Favorable conditions

» When there is a lot of rain just around the time that new growth and flowers are emerging, this disease tends to spread rapidly.

» Rainfall, high humidity, intermittent showers, and a temperature regime of 24 - 32°C during the flowering and fruit-setting stages of development are all conducive to the onset and spread of this disease.

» Warm, damp weather is ideal for the spread of the disease.

Management

» The trees in an orchard should not be too close together, and proper phytosanitary precautions should be implemented.

» When an infection first appears, it is best to cut off and burn off any infected tissue. Replace any limbs or branches that have died. That's because they're the main cause of the disease.

» Copper oxychloride, copper hydroxide, zinc and manganese carbamate, captafol, benomyl, dithiin, anilazine, bitertanol, triadimenol, and triforine can all be sprayed on plants to kill pests.

» Spray a 1% Bordeaux mixture while the plants are in their fresh flush or blooming phases.

Gummosis, *Lasiodiplodia theobromae*

Symptoms

» Trees of all ages were impacted.

» The liquid that initially seeps out of the main stem and branches turns black due to the disease. Once the leaves begin to yellow and fall, this is the first indicator that everyone will notice.

» On the affected area, gum secretions also manifest as longitudinal fissures.

» A reddish brown colour with tiny cavities filled with a red fluid will characterise the afflected tissues on the inside.

Management

» To treat the infection, chisel off the damaged tissue and then either use Bordeaux paste or dab the area with a Copper oxychloride liquid.

- Thin out the infected branches and spray them with a combination of copper sulphate, benomyl, and chlorothalonil + methyl thiophanate.

Pink disease or die-back, *Erythricium salmonicolor*

Symptoms

- Mycelium or a white or pinkish growth on the bark of affected branches first becomes apparent.
- The shoots eventually wither and die as the fungus spreads deeper into the tissues, producing a pinkish growth that is actually spores.
- The bark cracks and peels off at later stages.
- A tree's bare appearance is caused by its leaves turning yellow and falling off.
- Drying occurs beginning at the tips of infected shoots.

Favorable conditions

- This disease seems to flourish in the wetter months.

Management

- This is accomplished by cutting off and discarding the infected limbs or shoots below the infection site.
- Applying Bordeaux paste (10%) to the cut areas to seal and protect them.
- The months of October and November are ideal for applying preventative sprays of Bordeaux mixture (1%) or another Copper fungicide.

Powdery mildew, *Oidium anacardii*

Symptoms

- Those living near the coast are at a greater risk of contracting this disease.
- The disease manifests as a whitish, fuzzy growth on young nuts, flowers, and leaves.
- Trees that have been badly impacted will have leaves that are wrinkled and dry, flower buds and blooms that wilt, and fruits that drop before they're ready.

Favorable conditions

- Rains during the blossoming season spread the disease more quickly.
- During the blossoming stage, high humidity and cool nights are ideal

conditions for the rapid growth and spread of the fungus responsible for the disease.

Survival and spread

- Between epidemics, *O. anacardii* persists as active infections on a few immature stems and flowers that bloom out of season.
- The pathogen overwinters on fully developed leaves high in the cashew tree canopy, where it causes annual outbreaks.
- Spores are spread mostly through wind.

Management

- Powdery mildew can be effectively treated with sprays containing sulphur and/ or potassium/sodium bicarbonate.
- Trees can be protected from the disease by applying Wettable Sulfur @ 0.2% twice, once before bloom and once during.
- Dust fine Sulfur twice on young trees, once before the flowers open and again after the fruit is set.
- Carbendazim sprays at a concentration of 1 gramme per litre of water can also be used to control the disease.
- More effective systemic fungicides have been recommended recently, including Bayfidan and Anvil.

Red rust, *Cephaleurus virescens*

Symptoms

- Leaf petioles, young twigs, and leaf surfaces will develop tiny, raised, round patches that are greenish grey in colour.
- The specks eventually merge into massive uneven and slightly elevated patches, changing colour to a reddish brown.
- Leaf spots are made up of several filaments that extend through the cuticle.
- Spots fade to a cream to white texture once spores are released.
- In this condition, photosynthetic activity decreases, leading to leaf drop.

Management

- Copper oxychlodide sprayed at a concentration of 3 milligm per litre of water effectively controls the disease.

Shoot rot and leaf fall, *Phytophthora nicotianae* var. *nicotianae*

Symptoms

» During the southwest monsoon, extensive leaf drop and shoot rot are seen (June - August).

» The stem initially develops black, elongated sores that ooze gum.

» In time, the infected tissue travels up the stem and down the leaves, eventually causing the tender parts of the plant to wilt and die.

» There are also black, elongated lesions on the mid rib of the mature lower leaves, which spread to the main lateral veins and the rest of the leaf blade.

» Diseased leaves fall off quickly.

Favorable conditions

» At the time of the south-west monsoon, the disease is at its worst.

Management

» The spread of disease can be stopped by spraying with a 1% Bordeaux mixture before the start of the monsoon.

References

Majune, D. J., Masawe, P. A., & Mbega, E. R. (2018). Status and management of cashew disease in Tanzania. International Journal of Environment, Agriculture and Biotechnology, 3(5), 264441.

Adejumo, T. O. (2005). Crop protection strategies for major diseases of cocoa, coffee and cashew in Nigeria. African Journal of Biotechnology, 4(2), 143-150.

Sundararaju, D. (2002). Pest and disease management of cashew in India. Cashew, 16(4), 32-38.

Muntala, A., Gyasi, S. K., Norshie, P. M., Larbi-Koranteng, S., Ackah, F. K., Ntiamoah, D. A., & Mohamed, M. A. (2021). Diseases and insect pests associated with cashew (Anacardium occidentale L.) orchards in Ghana. European Journal of Agriculture and Food Sciences, 3(5), 23-32.

Nair, K. P., & Nair, K. P. (2021). Cashew nut (Anacardium occidentale L.). Tree Crops: Harvesting Cash from the World's Important Cash Crops, 27-77.

7 Cassava

INSECT PESTS

Tobacco whitefly, *Bemisia tabaci*

Damage symptoms

» The plant's vitality is drained when whiteflies feed on the plant's sap.

» When these pests attack, leaves wilt and curl and develop yellow spots.

» In extreme cases of infestation, black sooty mould appears.

» White waxy and powdery scale covers the bodies of both adults and juveniles.

Pest identification

» The small eggs, which are around 0.25 mm in diameter and stand upright on the leaf surface, are oval in shape. After being laid, an egg's white hue fades and it becomes a tan colour.

» Larvae are 0.3-0.7 mm in length, translucent, and oval in shape.

» The pupae are covered in a sticky wax and honeydew and have a dingy white colour.

» New adults can be identified by their bright red eyes at this time.

» About 1 mm in length, adults have two sets of white wings and pale yellow bodies.

Management

» When it comes to controlling whiteflies, good field cleanliness is the most practical strategy.

- Controlling the potential for a carryover population also includes removing crop waste and rogueing infected plants.
- The use of one to four yellow sticky cards for every 300 square metres of playing space.
- Intercropping with maize and cowpea decreased the prevalence of cassava mosaic disease by 20% and resulted in a 50% reduction in the adult whitefly population.
- Nanma, a CTCRI bioformulation made from cassava plants, is an extremely efficient pesticide.
- Extract of Neem seeds, for use in a spray.
- Use 15 ml of Neem oil per litre of soapy water and spray it on.
- On cassava, the most common and promising parasitoid of *Bemisia tabaci* is *Encarsia transvena*.
- *Eretmocerus mundus* and *Encarsia sophia*, two aphelinid parasitoids, were discovered feeding on *Bemisia tabaci* on cassava.
- *Beauveria bassiana*, a fungal infection, preys on whiteflies in humid climates.
- Methods for growing resistant crops.

Spiraling whitefly, *Aleurodicus dispersus*

Damage symptoms

- Whiteflies, in their characteristic spiral formation, are found in large numbers on the undersides of the leaves.
- A trail of white wax follows the adult as she deposits her eggs in an uneven spiral pattern.
- Both juvenile and adult stages feed on the plant's sap, damaging the host and leaving behind unsightly yellow spots on the leaves.
- Leaves are turning yellow.
- Honeydew secreted by whiteflies covers severely affected leaves with black sooty mould, which in turn reduces photosynthesis.
- One sign of a strong infestation is an early leaf drop.
- Most of the bug population booms during the dry season.

Pest identification

- » Eggs range in colour from a light yellow to a tan, and they are oval in shape and only 0.3 mm in length.
- » A nymph covered with several, closely spaced, short rods of wax, similar to glass, that form the sides of her body.
- » The powdery or waxy scales found all over an adult's body and wings create a dazzling white appearance. Males typically outweigh females.

Spread

- » The spiral whitefly can spread both by actively flying to new locations and by hitching a ride on the stems of plants.

Management

- » The alternative host abutilon indicum must be eradicated from the field.
- » Whiteflies can be caught with the help of a yellow sticky trap.
- » Use of 0.05% Dimethoate 30 EC, Biomite, or Profenofos as a spray.
- » Extensive mortality in all life stages of the pest was also observed when Neem oil and Cotton seed oil were used at 0.01%.
- » Adult spiral whiteflies can also be repelled by using 2.5% insecticidal soap.
- » Colored sticky traps can be effectively utilised for managing this notorious insect.
- » Predators including *Cybocephalus* sp., *Cryptolaemus montrouzieri*, and *Mallada astur* can also be efficiently used for the biocontrol of spiralling whitefly.
- » *Encarsia haitiensis* and other parasites meritoria, and the *Eretmocerus* species. are extremely common there.

Stem mussel scale, *Aonidomytilus albus*

Damage symptoms

- » Both adults and juveniles feed on the plant's lower stems.
- » The bug gets its sustenance by sucking on cassava stalks. The stems lose too much moisture in this situation to be planted.
- » Serious issue with the field crop and with the storage of planting materials.
- » In a quick multiplication, the scales cover the entire stem.

Pest identification

- Females are covered in a white substance and lack wings. For males, having wings is an advantage.
- Gray elongate mature scales are 2.5–3.0 mm long.

Favorable conditions

- Infestations happen in dry conditions and worsen during persistent drought.

Spread

- Male cassava white scales have the ability to glide.
- Pest spreads largely by wind and the transfer and planting of contaminated stem cuttings.

Management

- It's important to utilise only clean, healthy cuttings when planting.
- Cutting off infected stalks.
- The cuttings can also be kept healthy by storing them upright in the shade.
- When used as directed, the CTCRI bioformulation "Nanma" is highly effective in combating this insect.
- Demeton-s-methyl 0.1%, Parathion 0.5%, Dimethoate 0.05%, Monocrotophos 0.05%, or Malathion 0.1% sprayed on the cuttings before storage, or dipped in it for 10 minutes, is advised.
- Many species of Coccinellids, such as *Chilocorus nigrita, Scymnus guimeti, Horniolus guimeti, Cheilomenes sexmaculata*, and *Stethorus parcepunctatus*, feed almost exclusively on scale insects.
- Two species of the Hymenopteran parasitoid *Aspidiophagus* prey on scale insects. combined with *Signiphora* species.
- Sree Harsha, Sree Apoorva, and Sree Atulya were discovered field resistant, while CO 1 and Indira Cassava were tolerant, and Sree Prabha demonstrated field tolerant reactivity.

White peach scale, *Pseudaulacaspis pentagona*

Damage symptoms

- The bottoms of stems are a common place to find this kind of scale.

- Particularly vulnerable are weak or elderly plants.
- Infestations of this scale, which subsists on the plant's sap, result in the yellowing and eventual death of the plant's leaves.
- Due of heavy infestations, growth may be stunted.

Pest identification

- Female and male circular papery scales can be found at the base of the stem and the leaves.

Management

- White peach scale can be eliminated by chopping off infected plant stems and burning them.
- Do not try to grow new plants from cuttings taken from plants that have white peach scale.
- Avoid planting new crops downwind from those affected with white peach scale, as the crawlers will move to the new crop through the wind.
- Spray the stems with soapy water (five tablespoons of detergent to four litres of water) or white oil (petroleum jelly). These sprays are lethal because they plug the air passages, suffocating the target.
- To combat these crawling insects, try synthetic pyrethroids like Lambda Cyhalothrin or Cypermethrin.
- White peach scale is prey for ladybird beetles and lacewings.
- *Encarsia diaspidicola*, a species of parasitoid wasp, has been distributed in Samoa and, more recently, Hawaii with positive effects.

Black scale, *Parasaissetia nigra*

Damage symptoms

- Dark brown to black domes cover the soft scales, which range in size from 2 to 4 millimetres, and form colonies on the stems.
- *P. nigra* is a pest that attacks young wood.
- Honeydew from the scales accumulates on the plant and adjacent surfaces, where it feeds a community of sooty moulds. Occasionally, ants will be drawn to the honeydew due to the sugary substance being secreted.
- Sometimes defoliation and stunting of new growth occur.

Management

- If scale insects are allowed to overgrow on cassava cuttings, the resulting plants will be of inferior quality or may not even sprout.
- It has been suggested that kerosene emulsion and Parathion be used to prevent soft scales.
- The soft scale insects, such the black scale, *P. nigra*, appear to be a favourite food of the orange-spotted ladybird, *Orcus australasiae*.
- Unlike other methods, the use of *Metaphycus helvolus* (Hymenoptera: Encyrtidae) was successful in eradicating *P. nigra* from several regions and reducing the population to commercially irrelevant levels.

Root scale, *Stictococcus vayssierrei*

Damage symptoms

- The cassava root scale is a pest that resides on the cassava plant's underground storage roots, feeder roots, and submerged stems.
- The insects caused growth stunting, leaf drop, wilting, and dieback at the plant tips in young plants.
- The storage roots of a cassava plant suffer diminution and deformation when attacked by root scale. When cassava was grown on the level ground, rather than the ridges, it was subjected to more severe insect infestations.
- When intercropped with other plants like maize and groundnuts, infestations on the cassava were far worse than when it was grown by itself.
- The scale insect's infestation on the tubers rendered them unsellable.

Pest identification

- The insects, which resemble ticks on cassava, are oval, reddish-purple or brown in appearance, and have a wing span of around 0.5 millimetres. There are no wings on them.

Spread

- *Anoplolepis tenella*, a type of ant, played a crucial role in spreading *S. vayssierei*.
- The importance of *Anoplolepsis tenella* has been proven. scale crawlers move around within and between host plants with the help of tenella.

Management

- A substantial decrease of ants was seen in a boric acid-sucrose mix. Six weeks is the target time frame for a tenella population boom.
- Use of the "Sadisa" genotype of bitter cassava can lessen the severity of root scale infection and damage.

Mealy bugs, *Phenacoccus marginatus*, *P. manihoti*, *P. herreni*

Damage symptoms

- Adults and juveniles inject a toxic toxin into the plant through the juice they suck from leaves and shoot tips, causing a variety of undesirable effects such as stunted growth, deformed leaves, premature leaf drop, and chlorosis.
- Cause honeydew, which sooty mould feeds on, to spread across the leaf, blocking out the sun and preventing photosynthesis. The harvest of tubers will suffer as a result of this.
- Causing shorter internode lengths and leaf clustering ("bunchy tops").
- Causes stem distortion, leaf drying, and ultimately defoliation of plants.

Pest identification

- Females have a body shape similar to an oval, but are pink in colour and covered in white waxy filaments; they are around 2.2 millimetres in length and 1.4 millimetres in width, and they lack wings while males have them.

Favorable conditions

- Damage is worse during the warm and dry seasons than during the wet.

Spread

- This pest lives on the outside of plant stems and is carried by the wind, water, rain, clothing, field equipment, animals, and people who are working in the field.

Management

- To avoid pests in your garden, plant from pest-free sets.
- Cassava sets should be soaked in Neem oil and soap solution for 5 minutes.
- The seriously affected plants should be burned.
- Use a mixture of Neem oil and Soap solution for spraying and repeat the

spray after 15 days.

- Use biopesticides "Shreya" and "Nanma" (developed at CTCRI, Trivanandram) (developed at CTCRI, Trivanandram).
- Mealy bug management is aided by eliminating ant populations.
- Use a 1 ml/L spray solution of Malathion or Fenitrothion.
- Some insecticides have been found to be effective against mealy bugs; these include Profenofos 50 EC 2 ml/L, Chlorpyrifos 20 EC 4 ml/L, and Dimethoate 30 EC 2 ml/L.
- *Cryptolaemus montrouzieri*, a predatory coccinellid, was able to keep mealy bugs under control in the warehouse.
- The parasitoids *Acerophagus papaya, Apoanygyrus lopezi* (= *Epidinocarsis*), and *Anomalicornia tenuicornis* were all shown to be highly successful.

Striped mealy bug, *Ferrisia virgata*

Damage symptoms

- Underside of leaves, around petioles, and stems are common places to find striped mealy insect.
- Sucking the sap, *F. virgata* infestations remain clustered around the terminal shoots, leaves, and fruits, causing the plants to yellow, wither, and dry up and eventually drop their leaves and fruits.
- Large amounts of sticky honeydew coat the foliage and fruit, providing a food source for black sooty moulds.
- A plant's photosynthetic surface area is diminished as a result of black mould and waxy deposits.

Pest identification

- It's a white mealy beetle with two dark stripes running lengthwise, long glassy wax threads, and two long tails.

Management

- During the early stages of a local outbreak, infected branches should be lopped off and burned.
- *Allograpta obliqua* and *Cryptolaemus montrouzieri*, two predatory coccinellids, have kept the population under control.

Hornworm, *Erinnyis ello*

Damage symptoms

- Cassava leaves turn brown because of hornworms.
- Loss of life and production in plants is possible if a severe infestation occurs within the first two months of growth.
- Processing costs are higher because the tubers are of lower quality, have a lower starch content, and are waterier.
- For food, larvae eat the plant material they hatch on.
- When larvae attack, they can eat away at the plant's leaves and ultimately ruin the harvest.

Pest identification

- The adult form of the larva has a whitish-green colour with spots of yellow across its body and a more obvious black and red spot on its abdominal segment.
- The forewings of an adult female are a lighter shade of grey than those of a male.
- Each of the back wings is red with dark grey streaks along the edges.

Management

- Squashing hornworms after being picked by hand is an efficient method of control.
- Polistes canadensis, a species of predatory wasp, is commonly used for effective management.
- The eggs of *E. ello* on cassava are killed by *Trichogramma fasciatum* at a rate of 60- 100%.
- The larval population of *E. ello* can be reduced by spraying the area with *Bacillus thuringiensis*.
- The hornworm larvae are susceptible to the *baculovirus erinnyis*.
- More than 2,000 hectares of cassava are being treated with granulosis virus (GV) to prevent the spread of *Erinnyis ello*.

Thrips, *Frankliniella williamsi*

Damage symptoms

» The upper and bottom surfaces of leaves are home to thrips, which are tiny, yellow, elongated insects.

» They drain the colour from leaves by sucking the sap, and both the young and the old do it.

» Young leaves become deformed if they are attacked whilst they are still developing.

» Spots occur on older tissues, making them look silvery or leathery and reducing their ability to produce oxygen through photosynthesis.

» Some species weaken plants by sucking the sap from growing tips, which in turn reduces productivity.

» There is substantial evidence that thrips can transmit serious viruses.

Favorable conditions

» The intensity of infestations increases as the dry season progresses.

Management

» Complete eradication of weeds and all traces of prior crop residue is essential for proper field cleanliness.

» Vegetables that are resistant to the insect can be grown as part of a crop rotation programme.

» In the dry season, bug populations can be reduced with the use of overhead irrigation.

» Cuttings can be prepared for planting by submerging them in hot water (a solution of equal parts boiling and cold water) for 5-10 minutes prior to planting.

» Horizontal, in-ground planting of the cuttings.

» Soaking the trimmings in an insecticide solution.

» Regent (Fipronil), Admire, Abamectin, and Vydate L are only some of the insecticides that can be sprayed to the upper and lower surfaces of the leaves, diluted in water at a rate of 5 ml to 4.5 L per litre.

Aphids, *Aphis gossypii*

Damage symptoms

- » Aphids are a common pest of cassava plants and can be seen in large numbers on the undersides of immature leaves, where they can cause deformation of the leaves.
- » When aphid infestations get bad enough, the leaves get chlorotic and eventually die.
- » The sap of immature plants is easily sucked by insects, which weakens the plant.
- » When infestation levels are severe, seedlings are weakened or died, and older plants that are infected have a slowdown in growth.
- » Pest-infected leaves will curl, shrivel, and even brown and die.
- » While consuming food, aphids excrete a sugary liquid called honeydew. Attracting ants and providing a food source for sooty mould (black fungus) both have negative effects on photosynthesis
- » Aphids can transmit diseases to other organisms.

Management

- » Proper field sanitation include clearing away unwanted vegetation and discarded plant material from earlier plantings.
- » Aphids are a favourite food for the ladybird beetle, a natural predator.
- » There is no need for chemical management of aphids when the aphid population is low and ladybird beetles are present.
- » If the population is too large, chemical methods of control may be used. Fastac, Decis, or Karate at 6 ml to 4.5 L of water; Sevin 85% W.P. (Carbaryl) at 6 g to 4.5 L of water; Malathion 57% EC at 15 ml to 4.5 L of water; any of these can be employed as contact or stomach insecticides.

Variegated grasshopper, *Zonocerus variegatus*

Damage symptoms

- » The leaves, petioles, and green stems of cassava plants are the primary food source for the variegated grasshopper.
- » Defoliation and stem debarking are two of the effects.
- » Grasshoppers, both adults and juveniles, work together to strip the cassava of its leaves and, in extreme cases, its bark.

- Cassava plants are more susceptible to pest damage in the dry season, and the damage is worse when the plants are older.
- The crop's output decreases after repeated defoliation.

Pest identification

- The typical adult has a dark green colour with yellow, black, and orange spots.
- Nymphs have yellow markings on their bodies, legs, antennae, and wing pads, and are otherwise black.
- The egg pods resemble the tiny groundnut pods.

Spread

- The grasshoppers with multiple colours and patterns are spreading from farm to farm. The insect, however, is not capable of long-distance flight.
- It spreads more rapidly in places where trees have been cut down than in forested regions.

Management

- The population of the pest can be controlled by destroying their egg pods.
- The *Chromolaena odorata* "Siam weed" is a breeding ground for the variegated grasshopper, hence it must be eradicated.
- To replace grasshopper larvae and adults on cassava, you can spray it with a "biopesticide" made from fungal spores suspended in oil.
- Spraying the product on *Chromolaena odorata* will kill the nymphs that congregate on the weed shortly after hatching.

Shoot fly, *Siblua pendula*

Damage symptoms

- Stunts the development of a plant by damaging its growing tips.
- Due to white larvae digging inside the developing point, it turns brown and distorted.
- A whitish-yellow fluid has begun to seep out of the broken shoot.
- It is possible for the growing tip to die and for secondary shoots to emerge, both of which are vulnerable to attack. At the start of the wet season, there is usually a large population explosion. Younger plants are more susceptible to damage than older ones.

Pest identification

- A mature fly has a metallic blue sheen.

Management

- Plants that are several years old or more are immune to economic losses caused by shoot fly.
- Young plants need to be protected with pesticides.
- There have been attempts to bait the adults with mixes of Malathion (0.15%) and Molasses (5%), however the larvae are said to be difficult to handle.
- The use of sprays containing diazinon, dicrotophos, dimethoate, methamitol, or monocrotophos has been suggested for the elimination of larvae.
- Planting of resistant varieties. Both the 14181AC and Ouro do Vale cultivars in Brazil have shown to be quite resistant to this insect.

Green mite, *Mononychellus tanajoa*

Damage symptoms

- Make a meal of the cassava plant's tender, underdeveloped leaves and green stalks.
- Underneath the leaves, along the veins and towards the leaf base, green mites can be found.
- Chlorotic spots (yellow spotting) emerge on leaves as a result of the loss of chlorophyll, which is caused by mites sucking fluids out of the cells of the plant tissue during an attack (green pigment).
- Leaves could turn spots and eventually fall off.
- Terminal (or newest) leaves die and fall off after a severe infestation by cassava green mite, leaving the stem tip looking like a "candle stick."

Pest identification

- Wings and antennae are absent. The maximum size for an adult female is 0.8 mm, making her slightly larger than a male of the same age.

Favorable conditions

- Generally speaking, the higher the temperature, the quicker the growth. The ideal temperature range is between 27 and 34 °C.
- If the drought lasts too long, the number of mites will rise, which would further diminish crop output.

Survival and spread

- Can last up to 60 days off of harvested leaves, stems, and cuttings.
- Propagated by infected wind gusts and plant parts (leaves, cuttings, root tubers, etc.).

Management

- Don't risk a plant failure by using dirty soil.
- Put seeds in the ground before the rains come.
- Double or even triple rows of pigeon peas can be planted as an intercrop.
- Bring in the predatory mites like *Typhlodromalus manihoti* and *Typhlodromalus aripo*, as well as the various predatory insects in the genera *Stethorus* and *Holobus* (=*Oligota*).
- Symptomatic leaves should be picked off the plant and thrown away.
- Destroy any crop remnants that may still have pests in them after harvest.

DISEASES

Root rot, *Macrophomina phaseolina, Botryodiplodia theobromae, Fusarium* spp.

Symptoms

- Damaged plants seem burnt and their leaves become dark and droop.
- It's possible for leaves to fall off the plant, but that's not the only option.
- It's inevitable that plants will die.
- If you look at the roots, you'll notice that they've died off and the tubers have grown.
- The tubers you buy at the store could be discoloured in a variety of shades, from light brown to dark grey to blue or pink.
- Roots that have rotted may be mushy and smell bad.
- White fungal formations at the stem base may appear during the wet season, indicating an infection.

Favorable conditions

- It's common knowledge that wet and poorly drained soils are breeding grounds for root rot diseases.

Survival and spread

» Root rot fungi can be found in the decaying matter of cassava plants on farms affected by the disease.

» Sources of the fungi that cause cassava root rot include polluted soils, cassava root, and stem debris.

» After using them in infected cassava fields, farmers should disinfect their tillage equipment to stop the spread of the fungus.

Management

» Cassava should be planted in drained soils, crop detritus should be burned, and tools should be cleaned after each use.

» Cassava thrives in sandy loamy soils with good drainage.

» Never plant cassava on soil that has a history of root rots or other serious diseases.

» In spite of seemingly healthy stems, planting materials shouldn't come from fields with obvious evidence of root rot infections.

» Fungi can be eliminated by burning plant detritus (stems and roots) after harvest.

» Rotating fields between cassava and grains can effectively lower inoculum (spores, etc.) levels.

» Rots are far less common on farms where harvests are completed early.

Tuber rot, *Phytophthora palmivora*

Losses

» As much as 80% of yield can be lost due to the disease.

Symptoms

» Effects plants of all ages, producing rapid degeneration and soft rotting of the bloated roots and premature death.

» When plants are infected, they produce rotting tubers.

» Without using any force on the ground, the plants may be pulled up simply.

» Lesions ranging in size from 15mm to 30mm in diameter, dark in colour, are observed on fully grown tubers in the wild.

» These wounds are surrounded by white mycelial mats produced by the fungus.

» Expansion of lesions leads to browning and oozing of tuber tissue from within.

» Depending on the soil conditions, infected tubers will let off an unpleasant odour and decompose within a week to seven days.

Survival

» Infected tubers that aren't used during the harvest are the primary means by which the disease survives in the soil.

» To a lesser extent, the fungus can also persist in the form of chlamydospores and latent mycelium.

Management

» Sanitation, ridge planting, irrigation regulation, drainage improvement, use of lighter soils, harvesting early, crop rotation with cereals, and fallowing are all practises that have proven successful.

» Plants with more space between them are less likely to contract a root disease.

» In order to prevent crop failure, crop leftovers must be collected and incinerated.

» Plants and seeds should be in good health before being planted.

» The *Trichoderma viride* (30 g/L for 15 minutes) treatment of the sett.

» There was a significant uptick in the number of marketable tubers and a decrease in the incidence of tuber rot when *T. viride* was applied to the soil at a rate of 1 kg/ha.

» It has been discovered that *Trichoderma viride, Trichoderma harzianum*, and *Pseudomonas fluorescens* can be used to prevent cassava tuber rot.

Brown leaf spot, *Cercosporidium henningsii*

Losses

» It can reduce tuber output by as much as 30 percent in vulnerable varieties, such as the Malayan-4.

Symptoms

» The leaves will develop distinctive brown dots (3-12 mm in diameter) on both sides.

» The spots on the top are evenly angular and brown, with a black border.

- When fungal fruiting bodies are present, spots on the underside of an object lose their sharp edges and take on a greyish hue in the centre.
- As the lesions grow, the leaf's borders and veins force them to take on a more uneven and angular form.
- In many cases, a halo-like border will emerge around the lesions.
- The disease causes the leaves to turn yellow, dry out, and eventually fall off.
- Severe defoliation of susceptible cultivars can occur during the wet season.

Favorable conditions

- *C. henningsii* thrives and spreads in humid, warm climates.

Survival and spread

- Old fungal infections, commonly found on dead leaves, help the fungus endure the dry season.
- The fungus's conidia, or spores, are carried by rain splash and wind to younger leaves and new plantings from older, infected leaves.

Management

- Wider spacing between plants and selecting cultivars with a low leaf area index are two recommendations for lowering humidity levels in a cassava field.
- If planting occurs while the plants are not in their most vulnerable growth stage (after 5 months old), the disease's impact will be mitigated.
- It was shown that spraying the disease with Copper fungicides such as Thiophanate, Benomyl, and Carbendazim once a month beginning in the third month was the most effective method of disease control.
- Both Sree Prakash and Sree Visakam, which originated in the Central Tuber Crops Research Institute in Thiruvananthapuram, Kerala, India, are resistant to brown leaf spot disease.

White leaf spot, *Cercospora caribaea*

Symptoms

- *C. caribaea* lesions are less severe and appear in a different colour than *C. henningsii* lesions.
- They range from being round to angular in shape, and can be anywhere from white to a yellowish brown in colour.

- Both sides of the lesions are sunken to the depth of a normal leaf blade.
- Lesions often have a broad coloured border on the bottom leaf surface, but the white spots themselves stay identifiable.
- Sometimes the border can look as a wavy, violet-brown line with a halo of yellow or brown.
- During pathogen fructification, which typically takes place on the underside of the leaf, the centres of the spots take on a greyish velvety appearance.

Survival and spread

- Fungi can overwinter in diseased, decaying tissue.
- There are other hosts for leaf spot fungus, and weeds are one of them.
- The conidia were dispersed by the wind and rain splash.

Management

- Sanitizing a field is clearing away fallen foliage, agricultural debris, and weeds from surrounding plants to stop the spread of disease.
- Rotate your crops in the same area.
- Planting media devoid of known diseases.
- If necessary, use a fungicide spray.
- Herbicides like Gramoxone, Roundup, and Karmex can be used to replace annual and perennial weeds that serve as alternate hosts.

Phomopsis blight, *Phomopsis* (*Diaporthe*) *manihotis*

Symptoms

- Large dark patches emerge with unclear edges.
- If you look closely, you can spot the lesions along the major veins or the midrib of the leaf.
- Upper surface of the lesions comprises of concentric rings generated by the brown pycnidia of the fungus.
- The underside of the lesions are uniformly brown, and just a few pycnidia are generated.
- High humidity reveals greyish-brown hyphae on the lesions' surface.
- When lesions form on the underside of a leaf, the veins and veinlets in that area turn necrotic.

» The lesions expand, causing leaf blight.

» Eventually the necrosis extends to the full leaf and petiole.

» The plants' leaves begin to wilt and fall, which can cause significant defoliation.

» Dieback of young branches can also be caused, and in extreme circumstances, the plant may perish.

» Symptoms of stem rot include a browning of the stem and the presence of pycnidia.

Survival and spread

» Transitional survival probably takes place as pycnidia or perithecia in dead leaves and stems between harvests.

» During the rainy season, when temperatures are often between 20 and 25 °C, diseases like malaria and dengue fever tend to spread more rapidly, whether they occur at a high altitude or in lowland areas.

» When water splashes over plant parts, it spreads the conidia that have grown on the leaves and stems.

Management

» It is possible to employ cultivars that have shown resistance in the field as a starting point for a breeding effort to create even more resistant varieties.

» Disease is no longer an issue thanks to meristem-tip cultivation.

Anthracnose, *Glomerella manihotis*

Symptoms

» Spots and leaf border blight manifest on newly emerging leaves, which also twist and curl.

» Necrotic infection causes tissue death; sometimes entire leaves are lost.

» This disease can also kill off young stems.

» Cankers (also known as "sores") form at the nodes where the petioles of leaves join the stems.

» Leaf drooping and wilting are symptoms of cankers, which weaken the petioles.

» Defoliation and shoot tip dieback or death of the entire shoot result from the wilted leaves dying and falling off.

- Fungal fruiting bodies, which are pink in colour, can be seen at the heart of the cankers.
- When the disease attacks cassava, it causes the soft sections of the stems to twist.

Favorable conditions

- In most cases, diseases emerge at the start of the rains and progress during the wet season.

Survival and spread

- If not removed following root harvest, the disease-causing fungus can survive in the dead stems and leaves of the cassava plant.
- A cassava plant infected with the anthracnose fungus is a likely source of the disease.
- The spores of the fungus are carried by the wind and land on healthy stems, or infected stem cuttings are planted elsewhere.

Management

- It is important to choose cuttings from healthy stems devoid of cankers and dieback tissues when propagating new plants.
- Stem cuttings must be soaked in appropriate fungicides.
- After harvest, diseased plant stems and leaves must be burned to prevent the spread of the disease.

Rust, *Uromyces manihotis*

Symptoms

- The pustules of this disease appear on the veins, petioles, or green stems.
- Depending on how old the pustule is or what kind of fungal fructification it has undergone, it may be orange or a light to dark brown colour.
- They typically cause the damaged plant to deform.

Management

- Aspersion of suspensions of *Darluca* spp. is proposed as a biological control method.

Bacterial blight, *Xanthomonas manihotis*

Symptoms

» Between the veins on the undersides of leaves, little, angular, brown, water-soaked sores have begun to develop.

» The affected area of the leaf turns brown as the lesion spreads.

» A golden halo may surround some lesions.

» Large necrotic patches emerge as lesions join together.

» Defoliation happens with leaf petioles staying in horizontal position while leaves drop.

» Failing shoots.

» The stems, leaves, and leaf petioles could have brown gum.

Favorable conditions

» Disease is more severe in wet environments.

Survival and spread

» The bacterium that causes the disease is spread from the stems of affected plants.

» Transmission via wet surfaces and contaminated implements.

» The exchange of contaminated plant cuttings is likely the most important vector of transmission.

Management

» Plants grown from meristem culture, rooted buds, or shoots are also excellent options for those looking to avoid the risk of disease.

» Stakes should be treated with either 50 °C hot water for around 50 minutes or a 10 ml solution of Cupric fungicides before being planted.

» Potassium, in particular, is applied at recommended rates, so plants thrive.

» Cassava and maize can be grown as an intercrop to lessen the impact of rain-splash and wind on the spread of disease.

» Crop rotation should include cereals, grains, or legumes in addition to cassava.

» Dispose of diseased plants and crop waste by burning them.

- Remove any diseased sections of the plant.
- In susceptible clones of cassava, foliar spraying of *Pseudomonas fluorescens* and *P. putida* dramatically decreased the amount of angular leaf spots per leaf.

Mosaic disease, *Cassava Mosaic Virus* (CMV)

Symptoms

- Leaves that have lost their green hue and instead are a mottled yellow or white may also be deformed and smaller in size.
- Plant growth can be slowed, leading to tuber splitting, poor root output, and poor quality stem cuttings in particularly susceptible varieties of cassava.
- Leaves with chlorosis discolouration in spots, which can be either yellow or green.
- The leaves range greatly in size and are frequently severely misshapen.
- Depending on the level of shrivelling, leaf blades may fold.

Spread

- The diseased plants are the primary vector for the disease's spread.
- Whiteflies, *Bemisia tabaci,* act as vectors for the secondary spread of the disease.

Management

- A excellent disease-free crop can be raised through careful selection of meristem- derived planting material, followed by clonal multiplication, screening, and rogueing of newly affected plants at regular intervals.
- Select plant materials that are free of diseases.
- Planting media free of viruses are produced in areas free of vectors (high altitude).
- Whitefly vector control using 0.05% Dimethoate spray.
- Up in the mountains, where the whitefly population is negligible, disease-free planting material can be multiplied at a rapid rate.
- Check your plants for signs of disease on a regular basis and dispose of or remove affected plants.
- The genuine article of cassava seeds being used.
- Rows of corn or sorghum planted along fences.

» *Bacillus subtilis* and *Enterobacter cloacae*, two rhizobacterial species, were found to significantly decrease CMV levels.

» Try planting Sree Padmanabha, Sree Vijaya, H-97, H 165, Sree Visakham, Sree Sahya, or any other cultivar that has proven to be resistant or tolerant in the field.

» Beads of consistent size, shape, and firmness, and a high in vitro germination rate were used to encapsulate nodes containing axillary buds from in vitro grown cassava variety H-226, resulting in synthetic seeds free of the cassava mosaic virus. The synthetic cassava seeds were stored in sterile (autoclaved) polythene bags under laminar flow at room temperature (25±21°C).

Brown streak disease, *Cassava Brown Streak Virus* (CBSV)

Losses

» Losses of up to 100% have been seen in some areas, while losses of up to 70% in vulnerable varieties are considered the norm.

Symptoms

» Vein banding that is chlorotic or necrotic in adult leaves, which may eventually join together to produce enormous yellow patches.

» Young stems have brown, elongated necrotic lesions.

» Tuber necrosis.

» Roots develop knots.

» Roots and tubers may develop a secondary fungal infection, which causes the internal tissues to turn brown and eventually rot.

» Lower and older leaves are turning yellow along the veins.

» The upper green part of the stem has dark brown dots on it.

» Symptoms include wilting leaves and a reduction in new growth in the affected area.

» When a tuber is cut open, the brown and hard rot that has set in can be seen.

» Roots become stunted and malformed as a result.

Survival and spread

» The virus that causes the disease typically spreads from plant to plant.

» Cuttings of infected stems are planted elsewhere, spreading the disease.

- The whitefly, *Bemisia tabaci* is the vector in the transmission of the disease.

Management

- When planting, only use cuttings from disease-free sources.
- Virus-free plants can be grown via thermotherapy and meristem-tip culture.
- To reduce the risk of brown streak virus, you should grow cassava varieties with higher resistance, such as Garukunsubire or Seruruseke.
- Any infected plants, or potential hosts, should be uprooted and destroyed.
- Necrosis of tubers can cause significant crop losses, so it's important to harvest early.
- To prevent the transmission of virus to a new place, it is important to observe plant quarantine procedures.
- Beneficial insects, such as spiders and ladybird beetles, can help reduce whitefly populations in the field.
- To keep track of whitefly populations, set up yellow sticky traps.
- Insecticidal soaps are sprayed under the leaf surface, where the insects breed, to eliminate the pest population.

References

Graziosi, I., Minato, N., Alvarez, E., Ngo, D. T., Hoat, T. X., Aye, T. M., ... & Wyckhuys, K. A. (2016). Emerging pests and diseases of South-east Asian cassava: a comprehensive evaluation of geographic priorities, management options and research needs. Pest Management Science, 72(6), 1071-1089.

Thresh, J. M., Fargette, D., & Otim-Nape, G. W. (1994). The viruses and virus diseases of cassava in Africa. African Crop Science Journal, 2(4), 459-478.

Bellotti, A., & van Schoonhoven, A. (1978). Mite and insect pests of cassava. Annual Review of Entomology, 23(1), 39-67.

Parsa, S., Medina, C., & Rodríguez, V. (2015). Sources of pest resistance in cassava. Crop Protection, 68, 79-84.

Bellotti, A. C., Smith, L., & Lapointe, S. L. (1999). Recent advances in cassava pest management. Annual review of entomology, 44(1), 343-370.

Night, G., Asiimwe, P., Gashaka, G., Nkezabahizi, D., Legg, J. P., Okao-Okuja, G., ... & Mutumwinka, M. (2011). Occurrence and distribution of cassava pests and diseases in Rwanda. Agriculture, ecosystems & environment, 140(3-4), 492-497.

8 Cinnamon

INSECT PESTS

Cinnamon butterfly, *Chilasa clytia*

Damage symptoms

- » Insect larvae eat young, yet maturing, leaves.
- » If an infestation is bad enough, the plant will lose all of its leaves except for the midribs and veins.

Pest identification

- » The eggs have a waxy appearance and an orange-yellow hue.
- » The larvae, which are resemblant to bird poop, are a pale yellow colour and have black patterns.

Management

- » If feasible, it is best to pull larvae and pupae off of trees by hand.
- » Spray Quinalphos at a rate of 2 ml/L on young and partially mature leaves to control severe infestations.
- » Parasitic egg worm *Telenomus remus* is set free.

Leaf miners, *Acrocercops* spp., *Phyllocnistis chrysophthalma*

Damage symptoms

- » Slender, grey larvae can be seen tunnelling their way into the leaf tissue.
- » They cause linear mines that finish in blister-like areas by feeding on the tissues between the upper and lower epidermis of young, delicate leaves.

- Crinkles and huge holes form infested leaves when mined areas dry up.
- Mined leaves become drab and curl, and new growth is stunted.
- Identified by their thin, white, meandering trails on leaves; excessive mining can cause blistering of leaves and the premature fall of leaves.
- The monsoon season is when leaf miners are most active, wreaking havoc on nurseries and primarily affecting young plants.

Pest identification

- The larvae start off a drab grey colour but develop into a pretty pink.
- Tiny, greyish and silvery when adult.

Management

- New flushes can be protected from infestation by spraying a 0.05% solution of Quinalphos at their appearance.

Common blue bottle, *Graphium sarpedon*

Damage symptoms

- At first, caterpillars only eat the undersides of cinnamon leaves, but subsequently they start to prefer the upper side's midrib for some reason.
- Due to their age-related transformation into a green colour, they were successfully hidden from view.

Pest identification

- When first laid, eggs are spherical, smooth, and a pale yellow in colour.
- The first-stage caterpillar has a lot of spines and a smoky appearance.

Management

- The elimination of diseased plant tissue.
- It's important to keep a close eye on the pest.

Psyllid, *Trioza cinnamomi*

Damage symptoms

- A jumping plant louse that causes galls on cinnamon leaves and is commonly found on the plant's foliage.
- The psyllid's eating triggers cellular abnormalities that manifest as galls.

» The psyllid being fed is trapped inside the growing gall.

Management

» The Trioza infestation that causes galls on shoots could be contained with regular trimming.

» Psyllid populations are typically controlled by natural predators.

» To control weeds, spray 30 ml/25 L of monocrotophos every 4 days.

» The homopteran psyllid can be managed by spraying either Quinalphos (0.05%) or Dimethoate (0.06%) three times at 15-day intervals during the vegetative bud stage. If necessary, you can spray again after 10-12 days.

Wood boring moth, *Ichneumoniptera xanthosoma*

Damage symptoms

» The larvae eat the soft tissue that lies between the bark and the woody stem, and they do this by boring into the base of the stems (cambium feeder).

» Pink stem borer damage, worm damage, and stem boring moth damage are all common names for what happens when these pests feast on a plant's stem.

» The caterpillar's and adult moth's ravenous appetite is to blame for the destruction.

» Small brown granules coming out of the caterpillar's feeding pores are its faeces.

» If the damage is severe enough, the plants may fall over from root rot and collapse.

» With no new growth coming from rotten stems, the clump could be wiped off in time.

Management

» Management of pests and diseases, as well as fertilisation and weeding, are essential for successful crop production.

» In order to limit damage to cinnamon by as much as 95%, it is recommended that the plant bases be regularly earthed up.

» The plants can be quickly treated by applying carbofuran to the soil around their roots.

» When used as a pesticide, sex pheromones (S3) [labelled as (E, Z)-3,

13-octadecadien-1] proved effective. The yellow Unitrap is the best trap and it should be placed at a height of 150 centimetres.

Gall mite, *Eriophyes dociersi* (*Aceria doctersi*)

Damage symptoms

» Galls, ranging in colour from pale green to yellow, appear on the leaf surface as a result of the nymphs feeding on the sap.

» Mites infest the galls that form on the upper surface of the leaf, which appear reddish at first and mature to a greenish colour.

» A significant amount of crop is lost as the gall dries and develops a dark brown hue after the adult emerges.

» Damage by microscopic mites that are difficult to notice with the human eye results in the formation of irregular galls on the undersides of leaves, leaf stalks, and/or new stems. Galls are roughly conical in shape.

Favorable conditions

» It was discovered that the number of new galls increases sharply following rainfall, coinciding with the flushing of new leaves.

Management

» Keep an eye out for gall mites by inspecting the leaf undersides on a regular basis; large trees usually suffer only minor harm from these pests, so management is unnecessary.

» The gall-bearing shoots caused by Eriophyes could be contained with regular pruning.

» Abamectin and Bifenthrin are two miticides with some evidence of success, but they must be sprayed before the mites enter the leaves and do their damage.

» The fewest amount of galls occurred after 30 ml Monocrotophos/ 25 L was applied every 4 days.

DISEASES

Leaf spot and dieback, *Colletotrichum gloeosporioides*

Symptoms

» Small brown spots appear on the leaves of early seedlings; the spots eventually consolidate into larger brown patches with an uneven shape; and the lesions themselves sometimes fall off the leaves, leaving a shot hole behind.

- Mild leaf spot disease causes partial wilting in these seedlings.
- The elder leaves, especially those in the shadow, often have tiny brown patches.
- These minuscule particles add up to huge necrotic patches that have a charred look.
- Later, the edges of the spots turn a dark brown papery texture.
- These lesions might start at the leaf's tip or spread outward from the leaf 's edges.
- In extreme cases of infection, lesions might span the width of an entire leaf.
- Dying-back signs develop as the infection moves up the stem after the leaves have dried up.
- The destruction of tender shoots.
- Necrotic spots appear as tiny specks on mature trees and eventually spread to cover large areas.
- Spots develop a concentric pattern of light to dark brown as wet and dry periods alternate.

Management

- The diseased plant matter needs to be gathered and disposed of.
- The removal of diseased branches is recommended.
- The severity of the disease may be mitigated by reducing shading, especially in the early phases of plant growth.
- Several different fungicides have been suggested, and they are as follows: There should be a 7-10 day gap between doses of chlorothalonil 75% WP, carbendazim 50% WP, mancozeb 80% WP, or thiophanate- methyl 70% WP.
- Apply a spray solution containing either 1% Bordeaux mixture, 0.3% Copper oxychloride, 0.2% Mancozeb, or 0.1% Carbendazim every 14 days to protect the crop.

Rough bark disease, *Phomopsis* spp.

Symptoms

- Affect plants of all ages negatively, but especially younger ones.
- Spots of black or brown colour occur on the stem bark, expanding into larger areas.
- These blotches have a border of dark brown or black around them.
- The affected bark became tough and scab-like looking.
- As the crop matures, these scabby patches spread throughout the bark, diminishing the quality of the harvest.
- Young leaves of damaged plants often show interveinal chlorosis.
- Extreme yield losses might occur as a result of the premature death of immature shoots.

Favorable conditions

- Disease is more common in areas with sandy soil, sloping ground, extended weeding intervals, and low doses of fertiliser.

Management

- Reduce the spread of disease by regularly removing and destroying infected bushes or seedlings.
- It's also possible to perform selective pruning.
- Disease rates can be lowered by spraying a solution containing 1-2% Bordeaux mixture.

Stripe canker, *Phytophthora cinnamomi*

Symptoms

- Stripe canker affects the wood of young trees, both the trunk and the branches.
- The stems have been observed to have developed vertical stripes, with amber exudates at the advancing borders that eventually solidify.
- Most of the dead bark stripes are found in the vertical direction close to the ground.
- Symptoms include sudden wilting or death of plants, broad strips of dead

bark growing vertically up trees, most often at ground level, and a thinning or loss of the tree's roots.

» Affected plants will have necrotic stripes that are both irregular and vertical (1-5 cm in width).
» The cankers are sunken and zoned off from the healthy tissue by a dark line.
» Causes plant death from a lack of water and nutrients by attacking and destroying root systems.
» Grass trees are especially vulnerable, but any patches of dead or decaying vegetation may be an indication of this quiet killer's existence.

Favorable conditions

» Unhealthy soils account for up to 42% of the damage caused by disease.
» The fungus flourishes between the temperatures of 24 and 28 °C.

Survival and spread

» Producing chlamydospores allows fungi to live for a long time in the soil.
» Disseminated by the movement of infected soil and gravel and diseased plants.

Management

» To avoid spreading disease, only plant seeds or use plant material that has not been contaminated.
» Keeping disease rates to a minimum is made easier by enhancing soil drainage.
» Tar for phytosanitization and wound dressing.
» Replace susceptible hosts with non-hosts for a minimum of four years.
» It's not a good idea to plant in soils that are too wet.
» Use a systemic fungicide to saturate the soil.
» A sulphur treatment to the soil has been suggested.

Pink disease, *Corticium salmonicolor*

Symptoms

» Initial signs include the appearance of pale pinkish white coated patches on stems or branches.
» The infection later spreads, killing the younger shoots after having destroyed the bark.

Management

- Infected plant trimmings and other debris must be burned to contain the spread of the disease.

References

Khan, A. U., Khan, A. U., Khanal, S., & Gyawali, S. (2020). Insect pests and diseases of cinnamon (Cinnamomum verum Presi.) and their management in agroforestry system: A review. Acta Entomology and Zoology, 1(2), 51-59.

Rajapakse, R. H. S., & Kumara, K. L. (2010). A review of identification and management of pests and diseases of Cinnamon (Cinnamomum zeylanicum Blume). Tropical Agricultural Research and Extension, 10.

Anandaraj, M., & Devasahayam, S. (2004). Pests and Diseases of Cinnamon and Cassia. Cinnamon and cassia: the genus Cinnamomum, 239.

Jayasinghe, G. G., Rajapakse, R. H. S., Kumara, K. W., Ratnasekera, D., & Adikaram, N. K. B. (2020). Pests and Diseases of Cinnamon (Cinnamomum zeylanicum Blume). Cinnamon: Botany, Agronomy, Chemistry and Industrial Applications, 201-232.

Devasahayam, S., Jacob, T. K., Eapen, S. J., & Senthil Kumar, C. M. (2022). Pests and Their Management in Spices: (Black Pepper, Cardamom, Ginger, Turmeric, Cinnamon, Clove, Nutmeg and Vanilla). Trends in Horticultural Entomology, 1331-1364.

9 Cocoa

INSECT PESTS

Mealy bug, *Planococcus minor*

Damage symptoms

- Waxy, disc-shaped insects with elongated bodies that hang from tree limbs.
- Ants, which may also spread the mealy bugs, are attracted to the insects.
- Sooty mould, which develops when fungi feed on the sugary honeydew that insects shed, may also be linked to the presence of insect colonies.
- Cocoa swelling shoot is a condition that could affect trees.
- Mealybugs feed on the sap of young pods and young stems.
- The first signs of damage are a slowing of growth and the appearance of extra branches at an unwelcome height.
- Cherelles eventually wilt and abort as a result.
- Spindle bugs, or mealy bugs, are a summertime problem, doing havoc on the plant's leaves, spathes, and bunches.
- The infected areas get yellow and dry out.

Management

- It takes two applications of 0.1% Fenthion or Monocrotophos to mealybugs.
- Use a 0.03% spray concentration of diazinon, dimethoate, or phosphamidon.
- Mealy bug infestations can also be treated locally using Monocrotophos (125 ml per 100 litres of water).

» The mealy bug was also preyed upon by two isolated endoparasitoids: *Coccidoxenoides perminutus* with *Leptomastix dactylopii.*

Aphid, *Toxoptera aurantii*

Damage symptoms

» Aphid adults and juveniles feed on plant nectar, and they thrive on the sugars in flower buds, new leaves, young stems, and tiny cherelles.

» Flower petals fall off and leaves curl as a result of an infestation.

Management

» Aphids have several natural predators that eat them and keep their numbers down. Syrphids, Cecidomyiids, Coccinellids, Hemerobiids, etc. fall within this category.

» Spraying Dimethoate at a concentration of 1.5 mg per millilitre in water is efficient for controlling aphids in extremely dense populations.

Stem borer, *Zeuzera coffeae*

Damage symptoms

» Caterpillars are responsible for creating unbranched hollow tunnels inside of the stem of immature branches.

» The symptoms are a round hole in the stem, the death of the tissue above the hole, and the scattering of faeces and fibres that have been eaten up.

» Young plants will not survive an attack on their primary stems.

» Only the branches are affected by the drying effects of an attack.

Pest identification

» The wingspan of a fully grown moth is between 30 and 50 mm. The thorax is white with four blue-black dots, and the wings and abdomen are also spotted.

Management

» Reduce the size of the tree or cut down and kill any diseased limbs.

» Apply a paste made of Carbaryl wettable powder to the affected area.

Pod borer, *Conogethes punctiferlis*

Damage symptoms

- » Damage to cocoa pod husk brought on by insect larvae entering and leaving the pod.
- » Pods ripening at different times and unevenly.
- » A pod-eating insect causes the seeds to clump together inside the pod.
- » If you've ever tried to separate cocoa beans from their pod after harvest, you know how tough it may be.
- » Damage to pods and flowers caused by a caterpillar.
- » Subsists on its own internal resources (pulp and seeds).
- » The pods wither and fall off before the seeds within mature.

Pest identification

- » The adult moth of this insect is tiny and brown.
- » Creamy in colour and measuring in at about 1 centimetre in length, larvae transform into a vibrant green upon emerging from their pods to begin the pupation process.

Management

- » Pest infested pods and blossoms are gathered and thrown away.
- » Using weed plants as alternate hosts for clean cultivation.
- » Fruits should ideally be wrapped in biodegradable plastics. Black ants and weaver ants are effective at controlling borer populations.
- » Observe adult activity with light traps spaced at 1/ha.
- » Application rates for Fenthion 1 ml, Endosulfan 2 ml, and Phosalone 2 ml/L.
- » Borers can be kept below economically destructive levels by spraying the underside of cocoa leaves with contact insecticides like pyrethroids or carbamates.

Mirids/Capsids, *Sahlbergella singularis*

Damage symptoms

- » Tiny puncture holes on immature stems and pods are the first sign of a problem.

» These wounds become necrotic very fast, leading to the appearance of black spots that may eventually evolve into cankers.
» Branches that have lost their natural hue.
» Leaves and branches at the ends of their lives.
» The trees stop bearing fruit.
» Pods, trunks, and twigs are staple foods for both nymphs and adults.
» Cocoa mirids cause unattractive necrotic lesions on the surface of cocoa stems, branches, and pods by penetrating the host cells there.
» Dying tips of branches and leaves are a common consequence of mirids feeding on shoots, which can lead to overall plant mortality.
» The wounds they leave behind are susceptible to infection by parasite fungus, which in turn can lead to branch and trunk death.

Pest identification

» The adult bug has lengthy legs and antennae and can be either red or brown. On average, an adult is 7–10 mm in length.

DISEASES

Black pod rot, *Phytophthora* spp.

Symptoms

» While infection of pods or cherelles (immature pods) can begin anywhere on their surface, it typically begins at the pod's tip or stem end.
» The lesion caused by the disease is solid and brown in colour; it spreads and eventually covers the entire pod.
» The infection of the husk may not reach the beans inside the pod for several days.
» Cacao beans become discoloured and shrivel when Phytophthora invades the internal pod tissues, as seen in more advanced infections.
» Infected pods turn black and mummy-like as they deteriorate.

Management

» Cocoa trees need enough of room between them so that the plantation may get plenty of fresh air.

- Each week during the rainy season, in particular, you should check for unhealthy pods and remove and burn them.
- Controlling shade by cutting back overgrown branches.
- Before the start of the monsoon season, you can protect your crops from disease by practising phytosanitation and spraying them with a prophylactic 1% Bordeaux mixture.
- Fungicide sprays containing copper as a preventative measure, used in tandem with systemic fungicides.

Witches' broom, *Crinipellis perniciosa*

Symptoms

- Typical characteristics include rapid growth from bud into shoots and brooms.
- Producing shoots that don't develop into fruiting limbs.
- Pods that are misshapen and have green spots on them appear to have ripened at different rates.
- Within flower cushions and on vegetative apical or auxiliary buds, colonised tissues s well and generate multiple succulent vegetative branches, known as brooms.
- Pods can also be infected by fungi, leading to necrotic lesions, irregular ripening, and other abnormalities.
- Pods and brooms that have been infected with a fungus turn progressively brown and dry.

Favorable conditions

- In order to sporulate, the pathogen needs wet and dry cycles, nighttime temperatures between 15 and 24°C, and relative humidities above 90%.

Spread

- Humidity has a significant role in the transmission of disease, with optimal conditions for outbreaks occurring at high temperatures and high humidity (>80%).

Management

- The practise of phytosanitation, in which infected plant portions are removed and destroyed, has been demonstrated to decrease pod loss and stall disease epidemics.

- Informed application of systemic fungicides like Cupric acid.
- *Trichoderma stromaticum* commercial formulations are presently employed for the control of witches' broom.
- Copper hydroxide and *T. stromaticum* together consistently produced more disease- free pods and healthier beans.

Frosty pod rot, *Moniliophthora roreri*

Symptoms

- Spots on the skin of developing pods.
- Dark brown spots that are quickly spreading across the pod's surface.
- Only pods in their reproductive stages are susceptible to the disease.
- Discolored, swollen pods are the first visible sign of infection, followed by the dense production of cream-colored spores, so-called frosty pods, on the pod surface within 2 weeks.
- Spore concentrations have been reported to be as high as 44 million conidia per square centimetre, and the sporulating area of the pods grows rapidly.

Spread

- The spores are powdery and dry, so they're easily dispersed by everything from the wind to the water to the pods themselves.
- It is at times of heavy precipitation that diseases spread the most.

Management

- The most cost-effective method of preventing the development of this disease is to remove the inoculum sources, which are dead sporulating pods, through regular and frequent harvests.
- If you want to protect your cocoa pods from disease, plant a kind that blooms during the dry season.
- Fungicides containing copper will help lower disease prevalence if they are applied.
- Spraying diseased pods with biocontrol agents (*Trichoderma* and *Clonostachys*) every two weeks to once a month has been demonstrated to decrease conidium formation.

Charcoal pod rot, *Botryodiplodia theobromae*

Symptoms

- Spots are pale yellow and originate primarily at the pod's tip or the stalk's end.
- Spots develop into dark brown, bigger lesions.
- The pods of infected trees turn black and mummify while still on the tree.
- Beans infected by the rot turn black and have soft, mushy tissue inside.

Management

- Take out all of the impacted pods.
- Mist the area with either the Bordeaux combination (1%) or Difolatan (0.3%).
- Insect and rodent pest management efforts will also contribute to lowering disease rates.

Swollen shoot virus

Symptoms

- Branch and twig enlargement.
- Chlorotic spots near the leaf veins.
- Spots or specks of chlorosis on the mottled leaves; smooth pods with less beans.
- Nodes, internodes, and the very tips of stems and branches can expand.
- Pods with a mottled pattern of hue.
- The xylem and phloem vasculature are enlarged.
- Typically, a thin crimson band will outline the veins on immature leaves.
- Pods have been seen with varying shades of light and dark green mottling.
- Pods with the disease are often smaller than their healthy counterparts and may only contain half as many beans as usual.
- Potentially fatal defoliation can spread from tree to tree over time.

Transmission

- Carriers include mealybugs.

Management

- In order to stop the spread of an infection, diseased trees and the trees within a certain radius of them (up to 5 metres if less than 10 trees are affected, and up to 15 metres if more than 100 trees are infected) should be cut down and destroyed.
- In order to prevent the spread of cocoa rust, cocoa plants should be separated by at least 10 metres, and non-host crops like oil palm could be used to provide a physical barrier between cocoa fields.
- Mealworms can be parasitized by the fungus *Aspergillus parasiticus.*

References

Baah, F., & Anchirinah, V. (2011). A review of Cocoa Research Institute of Ghana extension activities and the management of cocoa pests and diseases in Ghana. *American Journal of Social and Management Sciences*, *2*(1), 196-201.

Ndubuaku, T. C. N., & Asogwa, E. U. (2021). Strategies for the control of pests and diseases for sustainable cocoa production in Nigeria. *African Scientist*, *7*(4).

Konam, J., Namaliu, Y., Daniel, R., & Guest, D. (2008). Integrated pest and disease management for sustainable cocoa production: a training manual for farmers and extension workers. Australian Centre for International Agricultural Research (ACIAR).

Keane, P. J., & Putter, C. A. J. (Eds.). (1992). Cocoa pest and disease management in Southeast Asia and Australasia (Vol. 112). Food & Agriculture Org..

10 Coconut

INSECT PESTS

Rhinoceros beetle, *Oryctes rhinoceros*

Damage symptoms

- Diamond or 'V' shaped cuts can be seen in the damaged leaves.
- When a frond unfolds, you can see a succession of holes along its length.
- The termination of the reproductive phase in newly planted forests.
- Fibrous material found eaten near bore holes is a telltale sign of infestation.

Pest identification

- Oval, white eggs.
- The grub measures 9-10 cm in length and is stout, meaty, filthy white, curved (C-shaped), and topped with a brownish head.
- The full-grown beetle is around 5 cm in length and has a stocky build. A large horn extends dorsally from its skull. Males often have a longer horn than females.

Management

- To prevent the spread of pests, have all dead coconut trees in the garden disposed of by fire.
- You should search the manure pit and the affected trees for beetle eggs, grubs, and pupae, and then eliminate them.
- At least once every three months, spray a 0.1% solution of Carbaryl over

the grub breeding grounds to eradicate the pests.

» To prevent the spread of the pest, inject 5 x 1011 spores of the entomopathogenic fungus *Metarrhizium anisopliae* per cubic metre of manure pit.

» Every 45 days, insert three naphthalene balls into the interspaces between the sheaths of the three innermost crown leaves.

» Put 150 gm of Neem seed powder mixed with two parts sand (1:2) into the axils of the three innermost leaves on the crown.

» Inject 5 g of phorate into the innermost two leaf axils twice, at 6-month intervals, using perforated sachets.

» Placing one Rhinolure pheromone trap per hectare will suppress beetles.

» At each harvest, check the tree's crown thoroughly and use an iron hook to remove the adult beetle.

Red palm weevil, *Rhynchophorus ferrugineus*

Damage symptoms

» There are tiny bore holes all over, and some of the chewed fibrous material is sticking out.

» The presence of brown liquid oozing from these holes is indicative of an early stage pest infestation.

» There is yellowing and withering in the central shoot.

» Inside the trunk at the damaged area, there were several grubs, pupae, and adults of the bug.

Pest identification

» Oval, white eggs.

» A mature grub will be thick, meaty, and apodous.

» The thorax of this weevil is reddish brown with six black markings. The male is easily identifiable by his long, hairy snout.

Management

» It is important to remove and burn any damaged or decaying palm trees in a coconut grove to stop the spread of any pests.

» Stem and root damage, as well as leaf loss, should be avoided at all costs.

» Fill any cracks with cement and drill a hole in the stem just above the attack sites to inject 10 ml of Monocrotophos.

- Set up a pheromone trap with Ferrolure at a rate of 1 per hectare.
- The roots were given 10 ml of Monocrotophos and 10 ml of water in a polypropylene bag measuring 7 by 10 centimetres.
- Remove the injured plant sections and wipe with coal tar.

Black-headed caterpillar, *Opisina arenosella*

Damage symptoms

- Printed pamphlets with silken galleries and frass.
- Brochures are being dried in a massive fashion.
- In the event of a devastating attack, the entire plantation would appear incinerated from afar.

Pest identification

- Scale-like ridges of egg yolk.
- The caterpillar is a drab brown colour with a greenish brown body and a dark brown head.
- The adult moth ranges in hue from straw to ash grey and has uniformly pale, white upper wings.

Management

- The infected fronds should be cut and burned.
- Young trees should have the undersides of their leaves sprayed with Carbaryl 0.1% or Malathion 0.05% (1 ml/L).
- The monocrotophos 5 ml/palm trunk injection is also beneficial.
- Leash the parasitoid larvae *Goniozus nephantidis* at a density of 3,000 individuals per hectare (ha) every several months, preferably among coconut trees.
- Release of parasitoids in the crown area is discouraged because predators such as spiders and reduviid beetles will quickly wipe them off.

Scale, *Aspidiotus destructor*

Damage symptoms

- Leaves with faint yellow dots.
- Turning yellow from stem to leaf.
- Browning and early fall of the leaves.

Pest identification

- The adult insect has the shape of a flattened oval and is a reddish brown tint, making it look like a scale.

Management

- Trees with infestations are pruned and then burned.

White grub, *Leucopholis coneophora*

Damage symptoms

- Yellowing leaves.
- Cracking of the nut before its time.
- There is a lag in flowering.
- When the tree's foundation is excavated, white grubs are revealed.

Pest identification

- In its adult form, this beetle has a brown body and striated wings that only partially cover the abdomen.
- Malathion 5 D applied to the soil at a rate of 25 kg/ha just before planting.

Management

- Monsoon rains attract adult beetles, which can be collected and disposed of by targeting specific plants like neem, Ailanthus, and Acasia.
- After a rainstorm, plant twigs of Neem containing their leaves in coconut groves to lure and kill the adult bugs.
- Light trap or bonfire should be set up at 1/ha.

Eriophyid mites, *Aceria guerreronis*

Damage symptoms

» Underneath the perianth, little inverted triangular patches of white or yellow may appear.

» The husk's exterior is prone to developing cracks and splits throughout its length.

» Exudation of a sticky substance from the damaged area.

» Nuts with a hardened husk and a malformed kernel.

» The removal of buttons.

Management

» Apply fertiliser at the rates of 1.3 kilogm of urea, 2.0 kilogm of super phosphate, and 3.5 kilogm of muriate of potash per palm tree each year to achieve optimal nutrient management.

» Neem cake (5 kg) and organic manure (50 kg) applied per palm tree every year.

» Intercrop the area around your coconut grove with sun hemp, banana, and shelter belt with casurina to deter trespassers.

» Buttons and developing nuts on bunches should be sprayed with Triazophos 5 ml/L, Monocrotophos @ 2 ml/L, or Carbosulfan 2 ml/L.

» Watering roots with 15 ml of Monocrotophos, 15 ml of Triazophos, or 15 ml of Carbosulfan per litre.

» Nuts should be harvested prior to root feeding, and then again 45 days later.

» The most lethal fungal pathogen against coconut mites is *Hirsutella thompsonii*, which may kill off between 35% and 70% of the pests it infects.

DISEASES

Bud rot, *Phytophthora palmivora*

Symptoms

» The first visible indication is the yellowing and eventual death of one or two young leaves in the spindle.

» Close inspection reveals brown or black spots on the spindle leaves, followed by the plant's drooping and bending.

- Smelly, putrid decay of affected interior tissues; death of the spindle.
- Because the affected spindle's base has died, it can be simply removed.
- More recently, symptoms have been detected in younger leaves close to the spindle.
- Later, the inner leaves gradually die off, too, until only the outer whorl of fully developed leaves remains in the crown.
- The disease eventually kills the palm, and with it the expanding bud.

Favorable conditions

- Heavy precipitation, extremely high humidity (over 90%), cold temperatures (18–20°C), and tapper and Rhinoceros bug bites.

Survival and spread

- Mold can live on in the body as dormant mycelium in infected tissues and as chamydospores and oospores in crop wastes left in the soil.
- Sporangia and zoospores disperse disease through the air.
- The inoculum spreads from infected trees by rain, insects, and tappers.

Management

- Milder cases of the disease can be achieved through better drainage, wider spacing, and weed control.
- The use of a 1% Bordeaux mixture as a preventative spray on immature palms.
- Spindles exhibiting the first signs of infection, a yellowing of the leaves, should be removed and treated with 1% Bordeaux paste.
- It has been discovered that the pathogen growth can be stifled by the presence of certain bacteria, including *Bacillus subtilis, Bacillus macerans, Bacillus amyloliquifaciens*, and *Myrothecium roridum*.
- Tolerant is the hybrid Malayan Dwarf x West African Tall (Hybrid PB 12).

Stem bleeding, *Thielaviopsis paradoxa*

Symptoms

- The stem's longitudinal fissures are oozing a dark reddish brown liquid.
- As the exudates dry, they develop black encrustations bordered by a brownish orange colour.

- A number of dark brown spots have appeared on the lower trunk.
- Underneath the discoloration, the underlying tissues rot, becoming black and fibrous, and develop cavities.

Favorable conditions

- Lack of garden maintenance and attacks from *Diocalandra* and *Xyleborus* bugs; excessive irrigation or rainfall followed by drought; shallow loamy soils or laterite soil with clay or rock layer beneath the soil.

Survival and spread

- Fungal spores (called perithecia and conidia, respectively) can be found in decomposing plant matter and soil.
- Conidia carried by the wind are mostly responsible for the epidemic's rapid spread.
- The rain and irrigation water both play a role in the propagation of disease.
- The disease is spread in part by the beetles that feast on infected plants.

Management

- A chisel can be used to remove the diseased bark tissue, and then 5% Hexaconazole (5 ml/100 ml water), hot coal tar, or Bordeaux paste can be used to seal the wound and prevent further infection.
- Calixin or hexaconazole, 0.1%, soaked in 25 litres of water per tree once every two months.
- Applying the roots three times a year with 5% Carbendazim or 5% Calixin.
- When performing intercultivation, take care to avoid causing yourself any mechanical harm.
- Incorporate a biocontrol agent (*Trichoderma*) weighing 50 gm per palm once yearly between August and September.
- Keep the palm tree watered during the hot summer months.
- West Coast Coconut vs. Research has shown that tall people are better able to stop bleeding.

Basal stem rot / Thanjavur wilt, *Ganoderma lucidum*

Symptoms

- The outer leaves begin to yellow and fall off, exposing the spindle.
- Outer leaf fronds droop and hang limp for months before falling off the tree.
- A brownish-red, sticky fluid seeps out of the stem's base and can travel as high as 3 metres, browning and decaying inside the stem.
- The palms begin to wilt and eventually die when sporophores (brackets) appear at the base of the stem, above ground.
- The roots of dying trees become discoloured and decay badly.

Favorable conditions

- Trees cultivated in sandy loam and sandy soils are more susceptible to waterlogging after heavy rains, low soil moisture content during the summer, and damage from weevils and insects.

Survival and spread

- The fungus can be found in soil and can live there for a long period.
- Basidiospores in the soil are the principal vector of infection and cause damage to root systems.
- Water from rain and irrigation contributes to the proliferation of the fungus.

Management

- Dead palms and palms with advanced disease should be removed and destroyed.
- Trenches one-metre-deep and thirty centimetres wide can be dug to separate infected palms from healthy ones.
- In the summer, it is best to offer consistent basin irrigation rather than flood irrigation.
- Spread 50 kg of farmyard manure and 5 kilogm of *Trichoderma*- enhanced neem cake per palm tree annually.
- When you can, plant bananas between other crops.
- Apply 1% Bordeaux mixture to the soil three times a year at a rate of 40 litres per palm tree.
- Quarterly root feeding with either 2 g Aureofungin-sol + 1 g $CuSO_4$ in

100 ml of water, 2 ml of Hexaconazole, or 2 ml of Calixin. Only early cases of the disease will benefit from it.

Grey leaf blight, *Pestalotia palmarum*

Symptoms

- At first, symptoms appear exclusively on the outer whorl of leaves, especially in older leaves.
- The leaflets have tiny yellow dots in the centre of a grey border.
- The centres of the dots gradually become a bluish-gray colour, while the periphery and halo are brown and yellow.
- Necrotic grey patches form from a cluster of dots.
- The entire leaf blade dries out and shrinks, giving it a blighted or burned appearance.
- Numerous black acervuli, either globose or ovoid in shape, populate the leaf upper surface.

Favorable conditions

- Conditions such as poor drainage, a lack of potash, heavy rainfall over a prolonged period (more than four days), and gusty winds all contribute to poor crop yields.

Survival and spread

- Soil containing infected plant detritus is the fungus's permanent home.
- In this case, conidia in the air are responsible for the disease's transmission.

Management

- The rotting, falling leaves must be regularly swept up and burned.
- Add in a lot of potash.
- Fix the soil's drainage problems.
- Before the rains come, spray the crown with a 0.25 percent Copper oxychloride solution or a 1 percent Bordeaux combination.

Lethal yellowing/Root wilt disease, *Phytoplasma palmae*

Symptoms

- "Flaccidity" of leaves, wherein they curve unnaturally inwards like the ribs of mammals, is a key diagnostic sign.

- Also noticeable is the leaf necrosis and yellowing that occurs at the leaflet margins.
- Leaves wilt from the centre of the whorl outward, and buttons and immature nuts fall off.
- The mature nut is tiny, and the kernel within is paper thin.
- In later phases, the crown size of trees decreases and they cease to bear fruit.
- Roots display signs of decay, which progresses from the root's tip backwards.
- Cracks and blotches appear on older roots, and the once white cortex turns a brownish black as it dries up in flakes.
- The distal ends of leaflets in the central spindle and part of the immature leaves go dark and shrivel up as the first sign of disease.
- After some time, the damaged area of the leaf crumbles away, transforming the leaf into a fan.
- The decaying accelerates the palms' demise.

Favorable conditions

- The lace wing bug, *Stephanotis typica*, thrives on sandy and sandy loam soils, and severe flooding is common.

Spread and survival

- Dead or highly diseased plants are used as primary inoculum sources.

Management

- Remove any palms that are too diseased to be profitable, and then replace them with tolerant cvs. All the offspring of CGD x WCT or CDO x CGD crossings, as well as the local talls, Malayan Dwarfs, Maypans, and so on.
- To prevent the spread of vector-borne diseases, spray the leaves with 0.01% Monocroptophos.
- Put Oxytetracycline injections into your palms.
- Use an appropriate amount of fertiliser.
- Spread 50 kg of farmyard manure per palm tree annually.
- Plant green manure in the basin, then work it in when you fertilise.
- During the summer, water the palm tree once every four to six days, using 600 to 900 litres of water each basin. Make sure there's adequate drainage for when it rains.

» Grow fodder crops, cocoa, tapioca, elephant foot yam, yam, peanuts, and pineapples alongside your main crop for a minimum of three years.

References

Lever, R. J. A. W. (1969). Pests of the coconut palm (No. 7). Food & Agriculture Org..

Abhishek, T. S., & Dwivedi, S. A. (2021). Review on integrated pest management of coconut crop. Int. J. Entomol. Res, 6, 115-120.

Kumara, A. D. N. T., Chandrashekharaiah, M., Kandakoor, S. B., & Chakravarthy, A. K. (2015). Status and management of three major insect pests of coconut in the tropics and subtropics. New horizons in insect science: Towards sustainable pest management, 359-381.

Tzec-Simá, M., Félix, J. W., Granados-Alegría, M., Aparicio-Ortiz, M., Juárez-Monroy, D., Mayo-Ruiz, D., ... & Islas-Flores, I. (2022). Potential of Omics to Control Diseases and Pests in the Coconut Tree. Agronomy, 12(12), 3164.

Rajan, P., Mohan, C., & Josephrajkumar, A. (2010). Integrated pest management in coconut. Improving productivity and profitability in coconut farming, 325.

Mohan, C., Josephrajkumar, A., Prathibha, P. S., Sujithra, M., Sajan, J. V., & Anes, K. M. (2022). Pests and Their Management in Coconut. Trends in Horticultural Entomology, 1411-1439.

McCoy, R. E. (1976). Comparative epidemiology of the lethal yellowing, Kaincope, and cadang-cadang diseases of coconut palm. Plant Disease Reporter, 60(6), 498-502.

DebMandal, M., & Mandal, S. (2011). Coconut (Cocos nucifera L.: Arecaceae): in health promotion and disease prevention. Asian Pacific journal of tropical medicine, 4(3), 241-247.

Oropeza-Salín, C., Sáenz, L., Narvaez, M., Nic-Matos, G., Córdova, I., Myrie, W., ... & Ramos, E. (2020). Dealing with lethal yellowing and related diseases in coconut. Coconut biotechnology: towards the sustainability of the 'Tree of Life', 169-197.

11 Coffee

INSECT PESTS

White stem borer, *Xylotrechus quadripes*

Damage symptoms

- Coffee made from Arabica beans is the most popular choice.
- Plants with white stem borer infestations have raised ridges on their stems.
- Plants become wilted and yellow as a result.
- Adult beetles tunnel out of the ground and emerge from the ground surface.
- Grubs tunnel in all directions, devouring the stem and thick primary branches for food.
- Expectations that look like chalk fill the tunnels.

Management

- Preservation of ideal shading conditions.
- Infested plants must be tracked down and removed on a regular basis.
- The loose, scaly bark on the main stem and thick primaries is brushed away using coir gloves.
- Applying Neem oil (2-5%) at regular intervals of 1-2 weeks.
- Cases are tracked with the help of pheromone traps.
- Apply 10% lime on the stem.
- The parasitoid Apensia sp. was released en masse. facilitate efforts to control the white stem borer.

Berry borer, *Hypothenemus hampei*

Damage symptoms

» Only found in coffee seeds, where it eats and reproduces.

» *Coffea Arabica* and *Robusta* (*Coffea canephora*) are also vulnerable to this pest.

» The female gains entry by slicing an asymmetrical hole in the cap of the coffee berry.

» Endosperm that hardens or becomes rigid in berries is where pollination and fertilisation take place.

» Inside-the-berry feeding in the field is primarily to blame for the destruction.

» Only 1 of a berry's 2 beans should be harmed. Half or more of the infected berry crop may be destroyed.

» Five percent to ninety percent of the berries have been reported damaged.

Pest identification

» Eggs that look like beans.

» An adult female is between 1.5 and 1.9 mm in length and has a brownish black coloration. Males are often much smaller and can't fly.

Favorable conditions

» It was determined that 25-26 °C and 90-95% relative humidity are ideal for reproduction and growth.

» Infestation is most prevalent between 500 and 1000 metres in altitude.

» Emergence and growth are encouraged by the sporadic rain.

» When comparing open or less dense shade with dense shade, the incidence was observed to be higher under thick shade.

Survival and spread

» Females can live for up to two months underground in buried beans, and females that are able to move about can do so by eating unripe berries.

Management

» Making sure there's enough shade and drainage.

» Complete and timely crop harvesting with little damage to the crop.

» Putting the berries in hot water for 2 minutes or cold water for 48 hours to kill any pests.

- Daytime application of the castor oil–smeared polythene sheet over the coffee harvest pile to catch the beetles.
- Attractant trapping at a density of 70 traps per hectare is recommended for catching beetles (1:1 ratio of ethanol to methanol works best).
- The months of September and October will see Endosulfan at a 0.07% spraying rate.
- Pesticides like chlorpyrifos, quinalphos, methomyl, and malathion were sprayed.
- The parasitoids *Prorops nasuta, Cephalonomia stephanoderis, Phymastichus coffeae,* and *Heterospilus coffeae* are being set free.
- Applications of the entomopathogenic fungus *Beauvaria bassiana* (10^8 conidia/ml) at strategic intervals.

Shot hole borer, *Xylosandrus compactus*

Damage symptoms

- Robusta coffee has a major pest problem.
- Suckers, young primary branches, and tertiary twigs are all targets.
- The adult female makes a pinhead-sized bore on the internodal side of the lower twig.
- On the infected plant, you can see the withered or dry branches riddled with bullet holes.
- Damaged leaves drop early at the attack site.
- The leaves towards the end of the plant begin to droop and dry out.
- It's clear that there's a lot of tunnelling going on up in the branches.
- The larvae live in a tunnel 38 mm in length that is lined with the ambrosia fungus (*Monilia* sp.).

Management

- Branches should be cut back between 5 and 7 centimetres below the hole.
- During the summer, cut down and eliminate any suckers that aren't wanted or are infested.
- It is especially important to take care of your trees between the months of

April and May and September and December by giving them plenty of light shade, making sure they have enough of room to drain, and regularly cutting back and burning off any diseased branches or suckers.

» In the months of April–May and October–December, dab a thick emulsion of Lindane or Carbaryl onto the stem. Adult beetles also emerge, take flight, mate, and lay eggs during this time.

Cockchafer beetle, *Phyllophaga (Holotrichia) conferta*

Damage symptoms

» Especially in child care facilities, grubs can do significant harm to young plants by eating at their roots.

» It is during the monsoon when the cockchafer beetle is most active (July to September- November).

» Between the months of May and October/November, they harvest feeder roots for consumption. Worse in overgrown grassy and weedy areas.

Management

» Get some decent shade up there.

» Lindane 20 EC @ 750 ml in 200 L of water should be drenched into the soil around the plants' bases.

» Plant with either 10 gm of phosate or 10 gm of aldicarb.

» Applying Aldrin or Dieldrin at a rate of 5% to the soil.

Green scale, *Coccus viridis*

Damage symptoms

» The insects tend to congregate in places with particularly juicy shoots.

» The pest's life cycle occurs in close proximity to one another on the underside of the leaves.

» They cluster on the underside of the leaves, near the leaf's midrib and veins, where the sap is most concentrated.

» As a result of the infestation, the leaves curl and the weak branches droop.

» Mold grows on the honey dew that is released by the scales.

» A greater proportion of the damage is done to young plants.

- Bruised fruit falling to the ground.

Pest identification

- One egg is laid at a time, and it is a pale greenish hue along with an elongated oval shape.
- Nymphs, also known as juvenile green scales, are oval, flat, and a pale yellowish green in colour, and they have six short legs.
- Shiny and pale green as an adult, females have a black, irregular U-shaped internal marking that is clearly apparent on their dorsal side.

Management

- Spraying 0.03% Dimethoate, Diazinon, Quinalphos, Dicrotophos, or Monocrotophos after performing necessary pruning should resolve the issue.
- Eliminate the nests of the ants.
- Dusting the shrubs and trees with either 1.5% Ekalux or 5% Malathion will keep the ants at bay.
- Apply one of the following to the stems and soak the roots of mature plants: 200 litres of water or a newly made neutral Bordeaux combination, coupled with 25 ml of Ekalux, 1000 ml of Folithion, and 1000 ml of Lebaycid, and 1 kilogramme of Fish oil rosin soap diluted in 100 ml of Sandovit or Vettoplant.
- Quinalphos 2.5 ml/L water spray.

Shoot and root mealy bug, *Planococcus lilacinus, P. citri, P. perissopneum*

Damage symptoms

- The insect feeds on both young plants and their roots.
- As a result of the infestation, plants will yellow, wilt, and eventually die.
- Berry loss, sooty mould growth, and feeder root death are also indicators of damage.
- This pest is made more widely available thanks to the efforts of ants.
- From February through June, the bug is at its worst.
- Chlorotic leaf tissue develops, flower buds fail to open, and berry size decreases.
- Honey dew released by insects turns black with sooty mould. plants with the disease have their leaves turn black.

Management

- Eliminate the nests of the ants.
- Dusting the shrubs and trees with either 1.5% Ekalux or 5% Malathion will keep the ants at bay.
- It's important to remove the weeds that mealybugs feed on.
- Use Phorate or Dasanit 5 G for treatment.
- » Wet the soil around the roots of mature plants and spray them with one of the following: 1 kilogramme of Fish oil rosin soap and 100 ml of Sandovit or Vettoplant in 200 litres of water or freshly produced neutral Bordeaux mixture.
- Any systemic pesticide should be sprayed at twice the normal rate.
- The parasitoid *Leptomastix dactylopii* should be released at a rate of 10 beetles per plant to protect citrus trees from the pest *P. citri* or the Australian lady bird beetle, *Cryptolaemus montrouzieri.*

Hairy caterpillars, *Eupterote canaraica*

Damage symptoms

- Spread out on trees that provide shade and eventually make their way to Arabica coffee.
- Gatherings that happen in groups.
- Defoliate plants by feasting on their leaves.
- It can take plants two to three years to recover from severe damage.

Pest identification

- The caterpillar's coloration is quite light yellow.
- The wings of an adult of this moth are marked with two diagonal bands and zigzag lines.
- Males are often shorter than females.

Management

- Whenever feasible, you should gather caterpillars and kill them.
- Between the months of January and May, you should be collecting pupae to burn.
- From June through July, put up light traps to kill moths.

- Spray a mixture of 1 kilogramme of Fish oil rosin soap and either 25 ml of Ekalaux EC or 280 ml of Thiodan EC on caterpillars to destroy them.
- Spray any of the following insecticides in a volume of 200 L of water: Quinalphos 25 EC (320 ml), Methyl Parathion 50 EC (160 ml), or Fenitrothion 50 EC (240 ml), plus 100 ml of an agricultural wetting agent.

Thrips, *Scirtothrips bispinosus, Heliothrips haemorrhoidalis*

Damage symptoms

- These thrips are incredibly delicate little insects that lacerate leaves for their food.
- The leaves of infected plants develop abnormal creases and mottling.
- Extremely common all year, but worse in warm weather.
- The female ovipositor looks like a saw.
- The female lays her eggs within the leaf itself.
- Adults and larvae alike can cause localised scarring and sap sucking damage to young leaves by grasping and scraping the epidermis.

Management

- Do your best to keep the shade.
- Until the shade trees have grown in, only spray open fields for green bug (other than Lannate).
- Eliminate the weeds and grasses that provide a safe haven for thrips.
- We recommend using a spray containing 0.03% of either phosphamidon, dimethoate, methyl demeton, or diazinon.

DISEASES

Leaf rust - *Hemileia vastatrix*

Symptoms

- On the underside of diseased leaves, little pustules of a brownish colour develop.
- Eventually, the surface develops an orange-brown hue and begins to sprout a granular mass of spores.
- The host tissue darkens, gets black, and dries up as the infection spreads

and the spore mass fades to white.

- In extreme weather, leaves will fall off

Spread and survival

- The uredospores, which are dispersed by rain splash and wind, can travel up to 1.5 kilometres from a single lesion.
- It's also possible for spores to be transported great distances by many creatures (insects, birds).
- Uredospore germination requires the presence of water, and infection can only take place through the stomata on the underside of the leaf.

Management

- The raking up and burning of diseased and decaying leaves.
- It is recommended to use a 1% Bordeaux mixture or 0.05% Tridemefon before, during, and after the monsoon season.
- You can control mosquitoes during periods of unpredictable precipitation by spraying a solution of 2 gm of Tridemefon and 0.5 ml of Teepol per litre of water every two weeks.
- Robusta coffee is known for its durability.

Black rot, *Pellicularia koleroga*

Symptoms

- Affected leaves, young twigs, and berries become black and rot due to black rot.
- Fallen leaves become attached by slimy fungal threads.
- Leaves fall off and berries drop.
- Bushes infected by brown rot gradually lose their leaves, turn yellow, and eventually die.
- Roots with this fungus are weak and show undulating dark brown spots.
- The lower portion of the stem becomes spongy and mushy.

Favorable conditions

- No long dry spells, just steady rain.
- High relative humidity (RH) of 95% to 100%.

- Complete cover from the sky, a place for plants to hide from the sun, and fresh air in the valleys.
- Mist that occurs frequently or continuously during the monsoon is a contributing element to the spread of the disease.

Survival and spread

- Pathogen hyphae enter palisade tissue intercellularly via stomata on the underside of the leaf.
- The fungus often disseminates from leaf to leaf via contact with the vegetative mycelium.
- There was a rapid spread of the disease due to the presence of infected plant matter.
- Mycelium colonises twigs and remains there year-round.

Management

- The upkeep of adequate drainage, ventilation, and exposure to sunlight.
- Mycelial threads and contaminated plant parts must be removed and destroyed in the moist early stage.
- Just before the monsoon rains start and again when they stop, spray the leaves and growing berries with a Bordeaux mixture (1%).
- Fungicides such as Carbendazim, Carboxyl, and Captafol, which work systemically and preventatively, have been proven beneficial.
- Good control of brown root and red root infections was achieved by applying 2 kg of *T. harzianum* or *T. viride*/plant to the soil between June and October. Coat 100 kg of FYM with 2.5 kg of *Trichoderma* spp. wait until the next day. Twice yearly, evenly disperse this over plant basins.

Damping off / Collar rot – *Rhizocotonia solani*

Symptoms

- There was a slowing down both before and after the emergence.
- There is a rapid demise of coffee spots.
- The stem of the coffee plant is mushy and rotting.
- Post-emergence damping off is characterised by an infection of the collar region just above soil level, which ultimately results in the death of the seedlings.

Favorable conditions

» Too much moisture in the ground.

» Experiencing difficulties due to excessive shade (insufficient drying of soil).

» Extremely dense planting (too many plants in a small area).

Survival and spread

» This disease spreads through the soil.

Management

» When preparing a nursery or repotting, always use fresh soil.

» Take care not to overwater.

» Plant seeds at a distance of 25 mm in rows and 100 mm in spacing between rows.

» Benlate (Benomyl) or Captan or Copper oxychloride soil drenches

Brown leaf spot, *Cercospora coffeicola*

Symptoms

» Leaves may develop irregular lesions that seem like dots or have brown, grey, or white cores.

» Spots' greyish white centres are surrounded by a brown ring (0.2-0.6" in diameter).

» A bright yellow "halo" can sometimes be seen around lesions.

» Dark brown lesions with ashy cores first appear on berries. These lesions may be longitudinal or irregular or oval in shape.

» Green berries with necrotic patches on their surface grow to cover the majority of the fruit.

» The rind of dried fruits becomes extremely sticky.

» Effects include leaf loss and coffee berry spoilage.

Favorable conditions

» Too much moisture in the ground.

» Either too much or too little sunlight.

» Stale air that isn't moving around.

» The soil has insufficient nitrogen and potassium.

Spread and survival

» The conidia of this virus are carried by the wind and can infect new plants.

Management

» Keep half your area in the shadow.

» In most areas, diseases can be managed with a healthy fertility schedule.

» Clean up your environment and remove crop debris.

» Make sure the soil drains well.

» Planting coffee in partial shade (between 35% and 65%) is ideal.

» You should keep watering levels down.

» In June and in the latter half of August, spray a Bordeaux mixture diluted to 1%.

» Spray foliage with fungicide (0.2% Copper hydroxide) in areas with ideal circumstances for the spread of infection and disease.

Die back / Anthracnose – *Colletorichum coffeanum*

Symptoms

» Leaves develop greyish or round patches measuring a metre or two in diameter.

» Berries can develop little sunken areas of a dark tint.

» Turning brown is a natural process for beans.

» As well as this, die-back can occur.

Survival and spread

» The outer bark is the primary site of infection because the fungus lives there as a saprophyte on dead tissue.

» Most of its conidia are released into the water during the rainy season.

» Conidia can be dispersed by rain water seeping through the canopy or by rain splash dispersing them amongst trees.

» Conidia are typically transported over great distances by passive vectors like birds, machines, etc.

Management

- Take care to ensure that the root zone drains well.
- To avoid defoliation, spray a 0.5% Bordeaux mixture/0.25% Mancozeb solution.
- It's important to keep the area adequately shaded to prevent sunburn.

Bacterial blight, *Pseudomonas syringae* pv. *garcae*

Symptoms

- At the onset of infection, lesions develop along the water-soaked margins of leaves.
- Brown, dried leaves roll inward and ultimately fall off.
- Dry, dead leaves do not fall off the plant but instead remain attached.
- In severe situations, the head may also show symptoms on secondary or tertiary branches.
- The last bud at the end of a branch is frequently assaulted. The infection spreads downward, causing the plant to wither.
- Flours and pinpricks wither and turn black.

Management

- Remove diseased branches by sucking them or cutting them off (at least 5 cm from point of infection).
- Sprays containing copper should be used often during the rainy season. During rainy seasons, spray every two weeks, and then every three weeks when rainy seasons are over.

References

Waller, J. M., Bigger, M., & Hillocks, R. J. (Eds.). (2007). Coffee pests, diseases and their management. CABI.

Cerda, R., Avelino, J., Gary, C., Tixier, P., Lechevallier, E., & Allinne, C. (2017). Primary and secondary yield losses caused by pests and diseases: Assessment and modeling in coffee. PloS one, 12(1), e0169133.

Carvalho, C. F., Carvalho, S. M., & Souza, B. (2019). Coffee (pp. 277-291). Springer International Publishing.

Ribeyre, F., & Avelino, J. (2012). Impact of field pests and diseases on coffee quality. Specialty coffee: managing quality, 151-176.

Gaitán, A. L., Cristancho, M. A., Caicedo, B. L. C., Rivillas, C. A., & Gómez, G. C. (Eds.). (2015). Compendium of coffee diseases and pests. APS Press, The American Phytopathological Society.

12 Coriander

INSECT PESTS

Aphid, *Cavariella aegopodii*

Damage symptoms

- Soft-bodied insects that live on the underside of leaves and/or stems, typically appearing in shades of green or yellow.
- Leaves may yellow, necrotic areas may become twisted shapes, and growth of the plant may be inhibited if aphids are present in large numbers.
- Honeydew, a sugary secretion from aphids, promotes the development of sooty mould on plants.
- Quality and quantity of the crop are both impacted by an aphid infestation.

Pest identification

- Small, glossy black eggs.
- Nymphs resemble the larger but winged adults in appearance.
- The adults have two long antennae that look like horns and have a thin, delicate body measuring only 1 to 4 mm in length. There are two little horns (cornicles) on the back of most aphids.

Management

- As long as the aphid population is contained to a small number of leaves or shoots, the infestation can be managed through selective pruning.
- Before planting, make sure the transplants are free of aphids.

» Aphids can be discouraged from feeding on plants by using reflective mulches, such as silver coloured plastic.

» A powerful stream of water can be used to blast aphids off the leaves of hardy plants.

» The most effective way of control is the use of insecticidal soaps or oils, such as Neem oil, canola oil, or 5% NSKE.

» For efficient pest management, spraying with 0.1% or 0.07% Endosulfan or Dimethoate, Monocrotophos, or Chlorpyriphos at 10-day intervals is recommended.

» Honeybees play a crucial role in the pollination of coriander. Therefore, caution should be exercised in selecting a pesticide that does not harm honeybees, or harms them very slightly. While honeybee activity drops in the evening, it's safe to spray endosulfan.

» Both the aphid *Aphidius colemani* and the parasitic worm *Diaeretiella* sp.

Cut worms, *Agrotis* spp.

Damage symptoms

» In some cases, the soil line may be used as a cutting guide for early transplants or seedlings.

» If the infection doesn't happen until the fruit is ripe, it will chew random holes in the skin.

» The damaging larvae are nocturnal, emerging from their hiding places in the soil around the plants' bases or in the detritus of uprooted plants.

» The larvae eat the roots, causing the plants to wither and die.

» As early as the seedling stage, this insect begins feeding on the plants, causing severe damage and yield reduction.

Pest identification

» The length of a larva is between 2.5 and 5 centimetres, and when it is disturbed, it will curl up into a C shape.

» The adult has a sizeable 40 to 55 mm wing span. The entire forewing, but notably its outer two-thirds, is a consistent dark brown. A bean-shaped wing spot is joined distally by a tiny but prominent black dash, and the distal portion is characterised with a lighter irregular band. The veins on the back wings are a darker shade of scale, and the overall colour is a light grey to white.

Management

- After harvesting, or at the very least two weeks before planting, clean the soil of all plant remnants.
- Plants can be protected from larvae by using plastic or foil collars around the stems, covering the bottom three inches above the soil line and reaching a couple of inches into the soil.
- You should go out at night and select the larvae by hand.
- Put some diatomaceous earth around the plants' foundations (this creates a sharp barrier that will cut the insects if they try and crawl over it).
- The control of cut worms can be achieved with the use of 4% Endosulfan dust at 20-25 kg/ha or 0.1% Chlorpyriphos at the time of ploughing.
- A 50,000-per-hectare release of *Trichogramma* sp.
- *Apanteles marginiventris, Chelonus insularis, Hyposoter exigua*, and *Lespesia archippivora* are examples of parasitoids.
- *Hexamermis arvalis* is an example of an entomopathogenic nematode.
- *Broscus punctatus* and *Liogryllus bimaculatus* are among their predators.

Thrips, *Thrips tabaci*

Damage symptoms

- Newly emerging leaves are a thrips' favourite feeding ground because of the tender, immature plant tissue they contain.
- As the leaf expands, the thrips damage from before becomes more noticeable, creating empty spots on the leaf's surface.
- Affected leaves will develop silvery spots or stripes that will reflect light.
- When the damage is severe, these microscopic patches can cover a large portion of the leaf's surface, preventing the plant from producing enough photosynthesis- fueling oxygen.
- The entire plant can turn white or silver and die after a strong attack.
- Bulb maturity and growth may be accelerated in wounded plants.

Pest identification

- Very tiny (0.5 to 1.2 mm), with an elongated, elliptical body and slender, long legs; the eyes have a deeper colour. Baby thrips can't fly because their antennae are too small and they lack wings.

» The fully grown wings of adults measure up to 2 mm in length and include a single longitudinal vein with many hairs attached perpendicular to the vein, giving the appearance of a fringe of hairs. When in rest, an insect's wings are folded down its back.

Management

» Controlling the thrips population by simulating rain with overhead watering.

» Be sure to keep the plants well watered throughout the season.

Mite, *Tetranychus* sp.

Damage symptoms

» Mites are a common problem for coriander when the seeds are forming.

» The young flowers and leaves are where the infestation is at its worst.

» They get their nourishment by sucking on leaves.

» Eventually, the entire plant's leaves will become a yellow white.

» Low yields and shrivelled grains result from a lack of timely pest control.

Pest identification

» Nymphs tend to be a sunny yellow.

» The full-grown specimens are around 0.5 mm in length, metallic brown to black, and have 35 bright yellow legs. Their forelegs stand out from the other three pairs as being noticeably longer.

Management

» Spraying the crop with 0.02% Ethion is an effective way to prevent damage from mites. Taking preventative measures or intervening early in an infestation's development will yield the best results.

» Spraying with 0.07% Dicofol or 0.03% Phosphamidon is another effective method of eradication.

» Common predators include lacewings (*Mallada basalis* and *Chrysoperla zastrowi sillemi*), mites (*Amblyseius alstoniae, A. womersleyi, A. fallacies,* and *Phytoseiulus persimilis*), coccinellids (*Stethorus punctillum*), staphylinid beetles (*Oligota* spp -*Feltiella minuta*).

DISEASES

Wilt, *Fusarium oxysporum* f. sp. *corianderii*

Symptoms

- The tips begin to droop, and eventually the entire leaf dies off and withers.
- Root vascular system discoloration, some wilting, and a halt in plant development.
- The leaves turn from pink to yellow to dry out and fall off.
- It is common to find sterility in plants of this type.
- If any seeds do develop, they are delicate and underdeveloped.
- Total crop failure due to severe infection at an early stage.

Favorable conditions

- The virus thrives on soils with high levels of moisture and warmth.

Survival and spread

- Primary infection is initiated by inoculum already existing in the soil, making this a soil-borne disease.

Management

- Summer is the best time to undertake deep ploughing.
- Rotating crops every four to six years is another option.
- Once this disease has been identified in a plot, no harvesting of coriander for at least two years is recommended.
- Before planting, treat the seeds with Carbendazim or Thiram at a rate of 2 g/kg.
- Water at 122 °F for 25 minutes will sterilise the seed.
- Carbendazim treatment of 2 gm per kilogramme of seed.
- *Pseudomonas fluorescens* 10 g/kg seed treatment or *Trichoderma viride* 4 g/kg seed treatment.
- Propiconazole (0.15%) was sprayed twice, 15 days apart, on Days 30 and 45 following planting.
- Carbendazim 0.1% or Copper oxychloride 0.25 % soil drench.

» Five kg per hectare of *T. harzianum* applied to the soil.

Powdery mildew, *Erysiphe polygoni*

Losses

» Powdery mildew causes roughly 50% loss in yield as it affects flowers and seeds.

Symptoms

» Seen as little white circles on developing stems and leaves.

» Affected leaves become chlorotic, decreased in size and deformed.

» White powdery growth arises on leaves, petioles, buds, flower stalks and bracts during its main stage.

» Severe infections can cause blooms to become deformed.

» Seed development may not take place in affected plants due to this disease.

» Infection leads in yellowing and loss of leaves leading in diminution of output.

Favorable conditions

» The disease is widespread during dry weather and low temperatures.

» Disease emergence is promoted by high humidity and moderate temperatures.

» Infection is especially severe in shady places.

» Fungus can persist in plant detritus in the form of cleistothecia and spread large distances by air.

Survival and spread

» Fungus can persist in plant detritus in the form of cleistothecia and spread large distances by air.

Management

» Early plantings resulted in decreased overall disease severity ratings.

» Apply 20–25 kg of sulphur per hectare as a dusting.

» Spraying of Dinocap, Wettable Sulfur 0.25% or Karathane 0.05 % or Carbendazim 0.1% at 15 days interval.

Blossom blight, *Aurobasidium* sp.

Symptoms

» Blighted field may have scattered little spots of brown areas or complete field turning brown.

» When a crop turns brown, it's usually because its growth tips, blooms, or umbels have died.

» Emerging umbels carrying young blooms brown and die.

» Infected plants develop dark brown patches on their stems and leaves, and any new flower umbels that form die.

» In the later stages of the growing season, when plants are beginning to develop, it may be difficult to identify a damaged field other than through decreased seed production.

Favorable conditions

» Conditions ideal for the spread of disease are high humidity (over 70%), high heat (between 12 and 25°C), and rain at irregular intervals.

Survival and spread

» Conidia or mycelia allow the pathogen to live on in infected plant matter, weeds, or soil.

Management

» To lessen the likelihood that a disease may spread to other places, it is recommended that you use seed collected from a symptom-free field.

» Registration of Quadris for the management of coriander flower blight has been completed. Before the disease has a chance to spread, a single application (between 0.45 and 0.13 L/ha) is allowed.

Stem gall, *Protomyces macrosporus*

Symptoms

» Shape of coriander seeds vary due to effect of the disease.

» This disease causes galls to form on the leaves and stems of infected plants.

Management

» Seed treatment with 4 g Thiram and 2 g Bavistin/kg seed.

- Spray 0.1% Carbendazim when the symptoms start emerging and repeat the spraying at an interval of 20 days till the disease is totally managed.
- To improve growth and yield while decreasing stem gall intensity, *T. viride*, *T. viride* [2 g] + *Melia azedarach* leaves (20 g/kg soil), and *T. viride* [2 g + *Eucalyptus globulus* leaves (20 g/kg soil) were shown to be the most effective.

Bacterial leaf spot, *Pseudomonas syringae* pv. *coriandricola*

Symptoms

- Small water droplets develop between the veins of a leaf, eventually becoming larger, dark brown, and eventually black.
- Long, black lines may run down the stems.
- Browning and yellowing of the inflorescence indicates disease.
- Damaged fruit that has been drenched in water.
- Leaf spots caused by the pathogen can result in a total crop failure.

Favorable conditions

- Continuous rain dramatically aggravated bacterial leaf spot disease.
- Splashing water from rains and overhead spray watering considerably increases the risk of bacterial leaf spot.

Survival and spread

- This disease remains dormant in plant until 'stress' is applied to growing plants (e.g. frost, hail damage).
- Disease is transferred through diseased seed and can be disseminated by splashing irrigation water and rain.

Management

- Only disease-free seed or seed with infection levels lower than 0.2% should be used in cultivation.
- Stopping irrigation for two to four weeks after a disease has been spotted is recommended for crops that rely on water to survive.
- Avoid overhead watering.
- Do not work with plants when they are moist.
- Copper fungicide treatments provide some protection.

References

Rao, S., & Garg, G. (2020). Importance of Pest and Pathogen Control System With Special Emphasis on Coriander Crop on the Indian Subcontinent. Ethnopharmacological Investigation of Indian Spices, 242-252.

Horn, L. N., Mulima, E. P., & Fwanyanga, F. M. (2023). Coriander Cultivation and Agricultural Practices. In Handbook of Coriander (Coriandrum sativum) (pp. 11-20). CRC Press.

Nair, R. (2021). Integrated management of aphid (Hydaphis coriandri (Das) and powdery mildew in coriander. (pp. 22-30). CRC Press.

Meena, N. K., Lal, G., Meena, R. S., Harisha, C. B., & Meena, S. R. (2017). Pest scenario of coriander (Coriandrum sativum L.) and population dynamics in semi-arid region of Rajasthan. International Journal of Tropical Agriculture, 35(4), 779-783.

Das, D., & Baruah, M. (2010). Sustainable practices for pest and disease management of horticultural crops. Annals of Plant Protection Sciences, 18(2), 357-361.

13 Cumin

INSECT PESTS

Aphids, *Myzus persicae, Aphis gossypii*

Damage symptoms

» Aphids are pests that can cause significant harm to plants by puncturing them and drinking their juices.

» They are particularly destructive to tender plant parts like young leaves and branches.

» Damage manifests itself in the leaves' inability to expand fully and their subsequent reduction in size.

» Damage from an infestation can cause young plants to wither and die.

» Black sooty mould thrives in environments where aphids have secreted a sugary liquid. Its presence can prevent sunlight from reaching the leaf surface, reducing photosynthesis.

» It drains the life force from vulnerable areas, resulting in a lower harvest.

Spread

» With their wings, aphids may easily hop from plant to plant, where they may unwittingly spread viruses they have picked up from diseased foliage.

Management

» Cumin should be sown between October and December, with November being the optimal timing for maximising yield and minimising aphid infestation.

- Applying a spray containing either 0.03% dimethoate, 0.025% methyl demeton, or 0.04% monocrotophos has been shown to be effective in reducing aphid populations.
- Aphids in cumin could be effectively controlled with two sprays of Monocrotophos and Acephate.
- When used at a concentration of 0.07%, endosulfan poses no risk to honey bees and can be sprayed at night, when they are less active.
- Use NSKE (3%), Neem oil (2%), or Tobacco decoction (0.05%), or soap with fish oil rosin.
- There are a variety of parasitoids, such as *Lysiphlebus* sp., *Diaeretiella* sp., *Aphelinus* sp., and *Aphidius colemani*.
- Ladybird beetles (*Coccinella septempunctata*), lacewings, spiders, hover flies, and other insects are only some of the predators that may come across your path.
- The aphid population was reduced by 90-91% compared to the untreated control after soil treatment of Neem cake + Mustard cake @ 0.5 t/ha, foliar spray of Acetamiprid 0.005% at 60-70 DAS, and Imidacloprid 0.005% at 70-80 DAS.
- The aphid doesn't attack the Cumin cv. MC-43 very much.

Thrips, *Thrips tabaci*

Damage symptoms

- Underneath the leaves, thrips cause damage by sucking the leaves' juices.
- They are particularly destructive to tender plant parts like young leaves and branches.
- The leaves turn a bluish-black hue and curl downwards as a result.
- Young leaves wilt and dry out when there is a severe infestation.
- Disease-causing viruses can be transmitted by thrips.

Pest identification

- Larvae are similar to adults in appearance, down to the absence of wings, and in that they are long and slender in body coloration and shape. Short antennae and dark eyes are other distinguishing features.
- The length of an adult is just around 1.5 mm; it has a long, yellow and brown

body and two sets of fringed (hairy) wings. The antennae have 7 segments and the mouth parts resemble a beak.

Management

- Use NSKE (3%), Neem oil (2%), or Tobacco decoction (0.05%), or soap with fish oil rosin.
- *Ceranisus menes* is a type of parasitoid.
- Many insects and animals serve as predators, such as the predatory thrips, the minute pirate bug, the ladybird beetle, the lacewing, the mirid bug, the hover fly, and many others.

Tobacco caterpillar, *Spodoptera litura*

Damage symptoms

- The pest reduces crop output by damaging plant leaves.
- At first, caterpillars live in colonies, and their scraping of the chlorophyll content of leaf lamina causes the leaves to turn a dingy white.
- Later, they develop into ravenous feeders that eat holes in the leaves in an erratic pattern.
- Drill random holes in leaves, which will eventually skeletonize the plant, leaving only the veins and petioles.
- Severe defoliation.

Pest identification

- At maturity, a caterpillar will be between 35 and 40 mm long. It has a velvety black body with yellowish-green dorsal stripes, white lateral bands, and a dark band that looks like an incomplete ring on the head and tail.
- The moth is of a moderate size and sturdy build, with forewings that range in hue from light grey to dark brown and include wavy white crisscross lines. The tips of the wings are white, but there are brown spots down the edges.

Management

- Planting castor and other oviposition trap crops.
- An early spraying of 0.02% Phosphamidon can prevent this pest from damaging the crop.
- *Trichogramma* sp., *Tetrastichus* sp., *Telenomus* sp., *Bracon* sp., *Campoletis* sp.,

Chelonus *sp.*, *Ichneumon sp.*, *Carcelia sp.*, *etc.* are all parasitoids.

» The lacewing, ladybird beetle, spider, red ant, dragonfly, robber fly, reduviid bug, praying mantis, King crow, and many more birds and insects serve as predators.

DISEASES

Fusarium wilt, *Fusarium oxysporum* f. sp. *cumini*

Symptoms

» Plants show withering indications of disease both at the seedling and later growth stages.

» When a stem is severed longitudinally, the vascular bundles reveal a brownish tint.

» One spray of either Carbendazim (0.1%) or Benomyl (0.05%) in addition to a seed treatment of 3g/kg seed was found to be beneficial.

» The application of the *Trichderma harzianum* T2 isolate as a seed dressing reduced disease occurrence by 65 percent.

» Researchers discovered that inoculating cumin plants with *Glomus fasciculatum*, *G. mosseae*, and *Acaulospora laevis* decreased the prevalence of wilt.

» Comparison of Cumin and Gujarat Only the GC-1, RZ-19, RS-1, MC-43, and RZ 209 strains are immune.

Favorable conditions

» Wilt is more likely to occur in soil with temperatures between 54.7 and 57.0 °F in the morning and 61.0 and 66.7 °F in the evening.

» This pathogen thrives in light soils.

Survival and spread

» The fungus can spread from soil to seed, and then from seed to plant detritus, where it lives on as a saprophyte as mycelium and chlamydospores.

» Short-distance dispersal occurs by irrigation water, rain splash, wind, and cross- cultural practises.

» The fungus spreads to other plants in the area after the plant dies, infecting them by spores it releases.

Management

- The disease can be stopped with summertime ploughing or solarization.
- It is crucial to select a field with good drainage to prevent water from pooling.
- Wilt is more likely to occur if watered too frequently.
- Seeds should be healthy and free of any known diseases.
- Substitute wheat, sorghum, cluster beans, and mustard for one another in a crop rotation.
- Neem cake is effective in preventing the further spread of fungus and, by extension, disease.
- Before planting, 2 kg of active ingredient (a.i.) Carbofuran 3G is used per hectare.
- Thiophanate Methyl and Carbendazim, applied to seeds and watered into the soil, reduced the prevalence of wilt and enhanced yield.

Powdery mildew, *Erysiphe polygoni*

Symptoms

- The crop is most vulnerable during the overcast months of February and March, when flowers are developing.
- The powdery growth typically begins on the leaves and might spread to the whole of the succulent plant.
- The potential for stunted seed growth is especially high in extreme cases.

Favorable conditions

- Conidial germination and disease development are facilitated by cool high humid weather (20-25° C) or gloomy weather with high relative humidity (RH) > 80%.

Survival and spread

- Both the soil and the seeds can harbour disease.
- Conidia are airborne and can travel great distances by the wind or splashes of rain, but soil and seeds are the primary vectors of propagation.

Management

- As soon as signs are observed, the crop should be dusted with 300 mesh

Sulfur dust at a rate of 25 kg/ha to prevent further spread of the disease.

- The best fungicide was spray propiconazole (0.025%).
- In the early stages of the disease, spraying with wettable Sulfur or Dinocap (Karathane or Thiovit) at a rate of 20-25 g per 10 litres of water can be effective in preventing it from spreading. A second spray, if necessary, should be administered 15–20 days following the first.
- Starting 30 days after emergence, spray twice with Mancozeb (0.25%) and twice with *Trichoderma harzianum*, with a 10-day gap between each application.
- It was also observed that spraying Mancozeb (0.25%) twice, followed by spraying *Pseudomonas fluorescens* twice, was effective.
- Sulfur 40% WP at 1.4 kg/acre in 400 L of water; Sulfur 80% WG at 0.75-1.0 kg/ acre in 300-400 L of water; Sulfur 85% DP at 6-8 kg/acre; or Dinocap 48% EC at 120 ml/acre in 300 L of water; second spray after 30 days gap.
- These strains are immune to disease-1 (GC-1), RZ-19, and RS-1.

Alternaria blight, *Alternaria burnsii*

Symptoms

- Disease strikes at the flowering stage during hot, humid weather, preventing proper maturation of the seed.
- Spots appear on the leaves as a result.
- Once the seeds have dried out, they are easily winnowed away by the wind.
- The disease pressure on an early-planted crop is so great that its seed is worthless.

Favorable conditions

- The conditions for the rapid spread of disease are met in damp weather, with temperatures between 20 and 32 °C, high humidity, and gloomy skies.

Survival and spread

- Diseased and partially decomposed agricultural waste, weeds in the cucurbit family, and possibly the soil all provide ideal conditions for the overwintering of fungi as dormant mycelium. Fungal conidia can persist for months in warm, dry environments.
- Besides the air, clothing, tools, and other equipment can disseminate a pathogen, as can running and splashing water.

» The conidia produced by the spores' germination are then carried farther by the wind, splashing rain, tools, or workers, penetrating sensitive tissue either directly or indirectly.

Management

» It was discovered that a combination foliar spray of Tebuconazole and Azadirachtin was the most effective method of application.

» Starting 40 days after sowing, it is advised to treat the seeds and spray them with 0.2% Mancozeb (Dithane-M-45) four times, each time waiting 10 days between applications.

» Difenoconazole 25% EC @ 0.05% in 200 L of water, second spray after 15 days interval; mancozeb 75% WP @ 0.6-0.8 kg/ acre in 200 L of water; zinceb 75% WP @ 0.6-0.8 kg/ acre in 200 L of water; aureofungin SP 46.15 percent w/v at 0.02% in 300 litres of water; spray again after 30 days.

» Reduced blight PDI by 56-63% compared to untreated control after seed treatment with Tebucanazole @ 2.5 g/kg, soil application of Neem cake + Mustard cake @0.5 t/ha, soil drenching with Metalaxyl (0.1%) at 25-30 DAS, foliar spray with Mancozeb (0.2%) at 60 DAS and Propiconazole (0.05%) at 75 and 90 DAS, and foliar spray with Dino

» Only the GC-1, RZ-19, RS-1, MC-43, and RZ 209 strains are immune.

Damping off, *Pythium* sp.

Symptoms

» There are two stages of damping off: before and after the appearance of a phenomenon.

» Seeds rot in the pre-emergence period, killing out the seedlings before they even have a chance to emerge from the soil.

» Upon breaking the surface of the earth, the seedlings are susceptible to a soft water soaked lesion at the collar region, which causes constriction and ultimately leads the seedlings to topple over.

» All of the seedlings rot away and die, including the young radicals and plumules.

Favorable conditions

» Infection and disease development flourish in conditions when there is lots of moisture in the air and soil, lots of nitrogen fertiliser, lots of cloud cover, and temperatures that stay below 24 °C for a few days.

» Pathogenic damping-off is exacerbated by crowded seedlings, excessive rainfall, poor drainage, and an abundance of soil solutes.

Survival and spread

» Infectious diseases can be found in dirt.

» The pathogen can be dispersed primarily through contact with soil and water, with further conidiary transmission occurring via rain splash and wind

Management

» To lessen the number of soil-dwelling pests, try solarizing the soil by covering the beds with 45 gauge (0.45 mm) thick polythene for three weeks prior to planting.

» Two to five tonnes per hectare of organic amendment (mustard, castor, or neem cake) should be spread.

References

Kafi, M. (Ed.). (2006). Cumin (Cuminum Cyminum): production and processing. CRC Press.

Didwania, N. (2019). Diseases of cumin and their management. Diseases of medicinal and aromatic plants and their management, 339-352.

Mishra, V. K. (2020). Insect pests of cumin and their management. Management of Insect Pests in Vegetable Crops: Concepts and Approaches"(Vishwakarma, R. and Kumar, R. eds.), 73.

Chaudhary, V., & Tripathi, R. S. (2010). Incidence of rodent pests in cumin (Cuminum cyminum L.) and their management. ICAR.

14 Elephant Foot Yam

INSECT PESTS

Mealy bug, *Rhizoecus amorphophalli*

Damage symptoms

- Obtain cell juice from the tubers.
- Deformed, severely infested tubers are not used in restaurants or sold in grocery stores.
- Infested the tubers completely with a white powdery mealy substance, changing their appearance irreparably.
- The number of stink bugs increased by a factor of four to five during storage.
- A decrease in field establishment and crop growth after planting corms was observed, suggesting that the pest had an adverse effect on corm quality.
- The delayed germination and subsequent growth of infested seed materials was visible within the first two months after planting.
- Lower yields resulted from harvesting contaminated seed material.

Favorable conditions

- High temperatures and little humidity stimulate fast reproduction.

Management

- Seed corms should be pest-free.
- Mealy bug populations and related corm damage could be reduced by pre-storage treatments with salt (Na Cl) solution (1,000 ppm), cow urine, cow

dung slurry (2 kg cow dung in 1 L of water), and clay slurry (1 kg of clay in 1 L of water).

- When planted in the field, corms that had been treated with a salt solution showed increased emergence and healthy growth.
- When applied to mealy bug nymphs, a mixture of a few drops of soap solution and 0.5% cassava seed extract (CSE) is effective.
- At 0.1%, surfactant was also extremely hazardous to early instars, killing half of them.
- Treatment with a 2% concentration of CSE only killed the females, indicating that they were not as sensitive as the first instars.
- If the infestation is bad enough, a spray of 0.05 percent Dimethoate or Monocrotophos will do the trick.
- Mealy bugs were killed at a 100% rate by Methyl parathion (0.05%), Malathion (99%), Chlorpyrifos (99%), and Quinalfos (99%) after 1 day.
- Up to 10-20% of stored mealy bugs have been parasitized by the encyrtid parasitoid *Anomalicornia tenuicornis* (Hymenoptera).
- *Rhizoecus* spp. were also successfully predated upon by the coccinellid *Cryptolaemus montrouzieri* (2 beetles per 1 kg tuber). In about three to four days, these two bioagents eliminated the mealybug infestation.

Scale insect, *Aspidiella hartii*

Damage symptoms

- The tuber is covered in the pest after it multiplies in storage.
- The tubers lose quality, viability, and marketability when their sap is removed during trade.
- In addition to preventing or even killing the plant, a severe infestation of tubers can prevent it from resprouting.
- The viability and quality of the harvested tubers are negatively impacted by an infestation that spreads through the seed tubers.

Management

- It is recommended to use pest-free seed tubers and to dip seed tubers in 0.1% Dimethoate before to planting.

- Planting material dipped in 0.5% Monocrotophos is also effective against scale insects.
- Dousing yams in a potent insecticide made from pyrethroids, an organophosphate.
- Pests can be managed through storage in sawdust, rice husk, wood ash, or a paraffin wax coating.
- Infected tubers treated with either a 3% Cassava seed extract or a 0.05% Chlorpyriphos solution.

DISEASES

Collar rot, *Sclerotium rolfsii*

Symptoms

- The pseudo-stem develops a water-soaked patch at its collar, and eventually the entire plant dies from rot at the collar.
- The collar area of the diseased plant displayed a white mycelial mat and many sclerotia.
- As a result of damage to the collar area and nematode (*Meloidogyne incognita*) infestation of the roots, plants are at risk for collar rot disease when used in intercultural operations.
- Damages the crop by rotting the collar region of the stem, causing it to collapse and cause a significant drop in yield and quality.

Favorable conditions

- Collar injuries, poor drainage, and flooding all contribute to the spread of disease.

Management

- Prevention measures include using disease-free seeds, ripping out diseased plants, enhancing drainage, and mixing Neem cake into the soil.
- Collar rot was least common and mildest in crops planted on March 28.
- Paddy straw and other organic material mulched deeply.
- The collar rot disease was lessened by mulching with a polythene sheet.
- Vermiwash was used to treat the seeds and soak the soil.
- Vermiwash 90 and 120 DAP applied through spray and tilled into the soil.

» Vermicompost applications decreased collar rot occurrence by 10-30%.

» We are applying 0.2% Captan to the soil.

» Injecting *Trichoderma* sp. (Td-1) isolated from *S. rolfsii* parasitized plants into the soil.

» The incorporation of biocontrol agents, such as *Bacillus subtilis, Bacillus cereus, Providencia rettgeri,* and *Trichoderma asperellum*, into soil by the use of Vermicompost.

» Neem cake was used in conjunction with *Trichoderma viride, Trichoderma harzianum,* and *Pseudomonas fluorescens* as biological control agents.

» Soil application/drenching with 600 g/ha of *Trichoderma viride* and Carboxin after treating seed corms.

» *T. harzianum*/Captan (0.2%) soil amendment in combination with corms.

» *Trichoderma harzianum* plus *Pseudomonas fluorescens* strain (P.s.c.) and fermented yeast extract (FYM) treatment.

» Neem cake, Ridomil MZ, and *Trichoderma harzianum* application.

Tuber rot, *Phytophthora colocasiae, Sclerotium rolfsii, Botryodiplodia theobromae, Fusarium, Rhizopus* sp., *Erwinia carotovora*

Symptoms

» If infected seedlings or tubers are utilised to grow elephant foot yam, the crop could be devastated by tuber rot.

» The problem of tuber rot is made worse by injuries sustained during harvesting and shipping.

» Root-knot nematode infestation causes damage to roots and tubers. *Meloidogyne incognita* serves as a susceptibility factor for a wide variety of fungi to infect a host.

Management

» Chisel plough ploughing on the deepest of soils.

» Planting with only disease-free tubers.

» Planting material treated with cow dung slurry containing *Trichoderma viride* is highly successful in preventing further infestation.

» Clean up by digging up and incinerating diseased tubers.

- Constructing a ridge using plantings.
- We need adequate drainage and carefully measured irrigation.
- Carbendazim (0.05%) treatment of seed carmel and storage on a cement floor resulted in the lowest rates of carmel rotting (7.57%) and sprouting (9.60%) and the longest storage life possible (148.65 days).
- Where *Botryodiplodia theobromae, Phytophthora colocasiae,* and *Pythium splendens* predominated and caused significant storage losses, benomyl was effective.
- Corm rots brought on by *P. colocasiae* can be stopped with copper oxychloride.
- The disease incidence was reduced (19.73% compared to 65.91% in the control) and the tuber production was raised (23.95 t/ha compared to 7.81 t/ha) after a 12-hour treatment with a 0.05% Agrimycin dip for the seed tubers.
- Applying 100 ppm Agrimycin three times, 15 days apart, after the first symptom of disease considerably reduced disease incidence (24.65% as opposed to 50.88% in control) and boosted yield (36.25 t/ha as compared to 17.81 t/ha in control).
- To lessen the occurrence of tuber rot, researchers used a conidial mixture of two bioagents: *Trichoderma pseudokoningii* (efficient in lowering the mycelial growth) and *T. virens* (good coloniser of sclerotia).

Leaf blight/root rot, *Pythium helicoides*

Symptoms

- As the disease advanced, plants exhibited symptoms ranging from chlorotic and stunted to fully blighted.
- Root necrosis symptoms started at the root's tip and spread rapidly, killing the entire root. It was simple to peel away the cortex of severely damaged roots, exposing the vascular system beneath.

Favourable conditions

- The disease thrives in climates with warm temperatures (22–23°C), high humidity (85–100 percent), and regular precipitation.

Management

- It's best to get your plants in the ground before the monsoon season hits.

- Plants are kept alive and well by using only healthy materials.
- Seed tubers treated with a biocontrol agent, specifically *Trichoderma viride.*
- To kill the fungus, you can use a fungicide spray, such as Mancozeb (0.2%) or Metalaxyl (0.05%).
- Maximum disease suppression was achieved by combining *Trichoderma harzianum* and *Pseudomonas fluorescens.*
- An effective strategy for combating leaf blight was discovered to be the selection of healthy tubers, followed by cormel treatment with *Trichoderma viride* at 5 g/kg mixed with cow dung slurry prior to planting, followed by application of *T. viride* enriched compost prior to planting, followed by mulching with paddy straw after planting, and finally, spraying of CTCRI formulation at 60 and 90 DAP.

Mosaic disease, *Dasheen Mosaic Virus* (DMV)

Crop losses

- Uttar Pradesh farmers have reported yield losses of 30-35%.

Symptoms

- Elephant foot yam frequently displays symptoms including mosaicing, mottling, leaf puckering, rolling, leaf narrowing, distortion of leaf lamina, and stunting.
- The corms grown from mottled plants are noticeably smaller than those grown from plants with normal leaves.
- Symptoms come and go, and their manifestation can shift from day to day or with the seasons.

Spread

- Under an electron microscope, the virus was shown to have flexible, filamentous particles (750 × 12 nm in size).
- Mechanical inoculation and grafting are two further modes of virus transmission.
- The aphids *Aphis craccivora, Aphis gossypii, Myzus persicae,* and *Pentalonia nigronervosa* are the vectors responsible for the virus's transient, ephemeral spread.

Management

- The most effective method of warding off this disease is to use seeds and cuttings from plants that have not been infected with mosaic virus. Only tubers that have not been exposed to the mosaic virus should be planted.
- Utilizing micro-propagation methods, we are able to cultivate plants devoid of mosaic virus.
- Mosaic disease was drastically decreased by treating infected tubers with hot air at 55°C for 10 minutes prior to planting, and subsequent spraying with Monocrotophos (0.05%) or another broad-spectrum insecticide 60 and 90 days after planting led to a comparable improvement in tuber output.
- To stop the transmission of the virus from aphids, which act as vectors, it is necessary to spray systemic pesticides.

References

Jayaprakas, C. A., & Harish, E. R. (2022). Pests and Their Management in Minor Tuber Crops: (Chinese Potato, Yams, Taro, Elephant Foot Yam and Tannia). Trends in Horticultural Entomology, 1109-1137.

Pillai, K. S., Palaniswami, M. S., Rajamma, P., Mohandas, C., & Jayaprakas, C. A. (1993). Pest management in tuber crops. Indian Horticulture, 38(3), 20-23.

Kumari, R., Singh, P. P., & Rai, R. C. (2013). Integrated management of diseases of elephant foot yam (Amorphophallus paeoniifolius). AROIDS: Opportunities and Challenges, 240.

15 Fenugreek

INSECT PESTS

Aphids, *Aphis craccivora*

Damage symptoms

- Insect nymphs and adults feed on plant sap, stunting the development of foliage, stems, buds, and flowers.
- Leaves may become yellow, develop necrotic patches and/or the growth of new shoots may be hindered if aphid infestation is particularly bad.
- Honeydew, a sugary secretion from aphids, promotes the development of sooty mould on plants.

Pest identification

- The eggs are little, and they're glossy and dark.
- Smaller versions of the adult winged creatures, nymphs resemble them in appearance.
- Adults are tiny, soft-bodied insects measuring only 1–4 mm in length and sporting two long antennae that resemble horns. Two little horns, or cornicles, protrude from the back of the head and body in most aphids.

Management

- Examine the transplants for aphids before putting them in the ground.
- To remove an aphid infestation, just cut off the infected branches or leaves.
- Aphids can be discouraged from feeding on plants by using reflective mulches, such as silver coloured plastic.

- Spraying a forceful stream of water onto sturdy plants helps dislodge aphids from the leaves.
- Ladybugs and lacewings are important, therefore let's keep them around and even increase their numbers.
- The most effective means of prevention include insecticidal soaps and oils like Neem, Canola, or NSKE 5%.
- The aphid can be managed by spraying a solution of 3 ml/L of dimethoate, 2.5 ml/L of methyl demeton, or 4 ml/L of monocrotophos.
- *Aphidius colemani*, *Aphelinus* sp., and *Diaeretiella* sp. are all examples of parasitoids.
- Insects like the red ant, the big-eyed insect (*Geocoris* sp.), the earwig, the cecidomyiid fly, the dragon fly, the lacewing, the ladybird beetle, the syrphid fly, and so on are all predators.

Thrips, *Scirtothrips dorsalis*

Damage symptoms

- Infested leaves roll upwards, become crumbly, and fall off the plant.
- Buds that have been infected with the pests get dry and fall off.
- Light brown scarring might be seen on affected fruits.
- Infestations in the early stages of plant development cause the plant to develop abnormally, cease producing flowers, and prevent the development of fruits.

Pest identification

- The eggs are hyaline, round, and placed en masse within the leaf tissues of leaves and shoots.
- Nymphs are extremely little, frail, and pale.
- The wings of an adult are extensively fringed and the bird is slim and a pale brown colour.

Management

- Blue sticky traps or water pan traps at a rate of 5 per acre should be used.
- Predators include predatory mites, thrips, hover flies, mirid bugs, *Orius* (pirate bug), and other species of *Oligota* and *Orius*.

Spotted pod borer, *Maruca vitrata*

Damage symptoms

» Spotted pod borer larvae consume the seeds inside the pods they damage. Water seeps in via the entry holes and discolours any unscathed seeds that have survived.

» Larvae of the fenugreek pod borer can cause damage to plants by tunnelling in the stems and webing young leaves together. This kind of destruction is not typical of how it acts on other legumes.

Pest identification

» The moth lays eggs that are flattened and a pale cream colour.

» The larvae have a creamy colour with two rows of paired black spots down their back.

» It's common for these patterns to fade to a nearly invisible state in the final instar.

» Lengths of up to 18 mm have been recorded for larvae.

» The wingspan of an adult is 20–25 mm, and the body is very thin. Their dark forewings feature a white band that starts at the leading edge and extends halfway back. There's a white dot inside this ring, very close to the front. The undersides of the wings are primarily transparent white, with a splotchy brown border.

Management

» However, helpful insects are unlikely to manage numbers exceeding 10 per square metre, thus it is still important to use broad-spectrum pesticides before the flowers bloom.

» Take the usual mechanical and biological precautions.

» A few examples of parasitoids are the *Trichogramma* species, the Bracon species, the *Apanteles taragamae,* the Carcellia, etc.

» Spiders, red ants, reduviid bugs, and praying mantids are also potential predators.

Red spider mite, *Tetranychus urticae*

Damage symptoms

» Spider mites use their long, needle-like mouthparts to eat the contents of the leaves' cells.

» This causes the chlorophyll content of the leaves to decrease, which in turn causes the appearance of white or yellow spots.

» When infestations are really bad, leaves wither and fall off.

» In extreme cases, the mites will also generate webbing on the leaf surfaces. When the mite population is very high, they will "balloon," or gather into a large, spherical ball, from which they will be carried by the wind to nearby leaves or plants.

Pest identification

» The eggs are round, hyaline, and have a ruddy hue.

» In appearance, nymphs are tri-legged, light-brown creatures.

» The adults are smaller and redder than the children. The females of this species are pyriform in outline, with a brilliant crimson front and a dark purplish brown back.

Management

» Don't let any weeds take over the field.

» Remove and burn crop waste that may harbour pests.

» Use NSKE (5%).

» Predators consist of insects such as lacewings (*Mallada basalis* and *Chrysoperla carnea*), predatory mites (*Amblyseius alstoniae, A. womersleyi, A. fallacies*, and *Phytoseiulus persimilis*), predatory coccinellids (*Stethorus punctillum*), staphylinid beetles (*Oligota* spp.)

» *Beauveria bassiana* is an example of an entomopathogenic plant.

DISEASES

Cercospora leaf spot, *Cercospora traversiana*

Symptoms

» Depressed circular lesions with haloes of chlorosis on leaves.

- Dead spots on leaves.
- Pods with discoloured spots.

Favorable conditions

- Temperatures between 25 to 35 °C, with nighttime lows above 16.1 °C, and a relative humidity of 90 to 95 percent are ideal.
- Temperatures below 15 °C and leaf wetness periods below 11 hours significantly minimise or eliminate infection.

Survival and spread

- Pathogen can persist as desiccation-resistant pseudo-stromata or conidia in plant detritus.
- Transmission via contaminated seed.
- When conditions are right, new conidia form and are carried to developing leaves and plants by the wind or splashing rain.

Management

- Sanitize your field regularly.
- Apply a spray of copper oxychloride at a concentration of 0.3%.
- The fenugreek variety 'Rajendra Kanti,' which has golden yellow seeds, is resistant to Cercospora leaf spot and has the ability to yield 1.5-1.8 t/ha, which is 40-42% greater than the yield of the local variety.

Powdery mildew, *Erysiphe polygoni, Leveillula taurica*

Symptoms

- This disease typically manifests itself in the later stages of a crop, when pod development is already well underway, and causes significant damage.
- Over time, white powdery spots develop on the underside and upper surface of leaves and elsewhere on the plant, and they spread.
- Sometimes the undersides of leaves will have yellow dots.
- Fungus's mycelium, conidiophore, and conidia make up the powdery spots.

Favorable conditions

- Powdery mildew can grow and spread rapidly in damp environments. Powdery mildew is more common when the days are cool and the nights are damp because high relative humidity promotes spore development and low relative humidity promotes spore dispersal.

- Later in the crop's life cycle, when the weather is warm and the foliage is dry, mildew often develops.

Survival and spread

- The fungus overwinters by attaching itself to plant matter, such as dead leaves.

Management

- Plant the seeds no later than October's end.
- Avoid overcrowding by ensuring appropriate spacing between plants.
- Mildew can be treated using Neem oil or insecticidal soap.
- It is recommended to spray the crop with Wettable Sulfur (0.2%), Hexaconazole (0.05%), Dinocap (0.06%), or Difenoconazole (0.04%) at the first sign of disease and then every 10-14 days thereafter.
- Propiconazole 0.1%, Penconazole 0.05%, or Carbandazim 0.1% sprayed twice 15 days apart is indicated for ringworm.
- Neem cake (1.0 t/ha) applied to the soil, together with *Trichoderma viride* (12.5 kg/ ha) and Neem seed kernel extract (NSKE) (5.0%) applied as a spray.
- Make sure to use a hardy strain, like as the Prabha.

Root rot, *Rhizoctonia solani*

Symptoms

- Disease symptoms, including root rot and leaf yellowing, first show in plants that are 30 to 45 days old.
- On both sides of the leaflets, there are little, chlorotic spots that collect water. They grow larger, take on an asymmetrical shape, and change colour to a brownish grey with a darker brown border. The middle eventually turned white, dried up, and crumbled away.
- When the growing tip of tiny lateral roots becomes infected with Rhizoctonia, root rot results. From there, the fungus spreads outward from the root tip and has the potential to cause the main root to rot. If a rootlet is infected with Rhizoctonia, the resulting root will typically have a "spear-point" appearance.
- Symptoms of this fungus include yellowing of lower leaves, withering, and slow growth.

» Once the root rot sets in, the plant will die.

Favorable conditions

» Conditions conducive to the spread of disease include high humidity and warm temperatures.

» Shaded areas with limited ventilation are ideal breeding grounds for disease.

Survival and spread

» The fungal spores, or sclerotia, live in the soil and can infect future generations.

» They may live in the soil for a long time without a host, and the inoculum levels there gradually rise over the course of several years (crop cycles).

Management

» Do a crop rotation every three to four years.

» Do not water too much.

» For optimum water drainage, use raised beds that are at least 15 centimetres high.

» In order to lessen the likelihood of rot, it's important to plant in soil that drains properly and is warm enough.

» Reducing root rot can be accomplished through amending the soil with Neem cake (1 t/ha) alone or in combination with *Trichoderma.*

» Carbendazim (0.2%) applied to the seedlings followed by a 0.1 percent drenching solution successfully manages this disease.

» *Trichoderma viride* seed treatment, 4 g/kg.

» Copper oxychloride 2 g/L or Carbendazim 0.5 g/L drenching.

References

Petropoulos, G. A. (Ed.). (2002). Fenugreek: the genus Trigonella. CRC Press.

Chhata, L. K., Ram, J., Qureshi, Q. G., & Maliwal, P. L. (2010). Integrated pest and disease management through organic farming approaches in fenugreek. Journal of Progressive Agriculture, 1(1), 45-47.

Acharya, S. N., Thomas, J. E., Prasad, R., & Basu, S. K. (2010). 19 Diseases of Fenugreek (*Trigonella foenum-graecum* L.) and Their Control Measures, with Specia l Emphasis on Fungal Diseases. Management of fungal plant pathogens, 245.

Ghule, M. R., Ramteke, P. K., Ramteke, S. D., & Jambhekar, H. (2021). Various Diseases Incidence on Fenugreek Crops and their Management. Fenugreek: Biology and Applications, 119-135.

Rao, A. B., Rajasekhar, P., Rao, G. R., & Rao, V. S. (2006). Seasonal incidence and management of Spodoptera litura on fenugreek. Annals of Plant Protection Sciences, 14(1), 86-89.

16 Ginger

INSECT PESTS

Shoot borer, *Dichocrocis punctiferalis*

Damage symptoms

- The larvae eat the plant's interior tissues, which causes the leaves to become yellow and wither once they drill into the false stems.
- The pseudo stem has an opening via which frass can be expelled.
- Damage from pests is often indicated by a centre shoot that has wilted and turned yellow.

Identification of pest

- Eggs are pink, oval, and round.
- Mature larvae are hardly hairy and a light brown colour.
- The wings of this medium-sized moth, which is about 20 mm across, are a bright orange-yellow and are speckled with tiny black spots when the moth is an adult.

Favorable conditions

- Humidity between 60 and 90 percent and 30 and 33°C.
- Insect and other pest populations peak in the field in late September and early October.
- Between July and October, the pest is at its peak of activity.

Management

» Using a spray bottle, apply either 2–5 ml/L of Nimbecidine or 2–5 ml/L of *Beauveria bassiana.*

» Apply Malathion 0.1% monthly between July and October.

» Shoot borer larvae were killed off completely by infections with *Steinernema* sp. (IISR 02) and *Oscheius* sp. (IISR 07 and 08).

Rhizome scale, *Aspidiotus hartii*

Damage symptoms

» When rhizomes are in the field or in storage, they are frequently infested by both nymphs and adults.

» In the field, infected plants grow weak, pale, and withering, which causes the rhizomes and buds to shrink.

» Dessication prevents germination of rhizomes.

» White-colored scales can be seen dispersed across rhizomes during the early stages of infestation, and later they cluster close to the developing buds.

» When the infestation is particularly bad, the rhizome and buds wither and the entire rhizome eventually dries up.

Identification of pest

» The females' scales are encrustations on the rhizomes, round (approximately 1mm in diameter), and a pale brown to grey colour.

» The male has a bright orange body, clear wings, and a well-defined skull, chest, and abdomen.

Favorable conditions

» Lacking precipitation

Management

» When storing rhizomes, avoid those that have been extensively infested. Drop them.

» Collect the fallen leaves and throw them away.

» For seed production, choose robust rhizomes that have not been infected with scale.

» In the case of a persistent infestation, treat seed material with 0.075% Quinalphos (for 20-30 minutes) prior to storage and planting.

» For 15 minutes prior to storage, submerge the rhizomes in a solution of 0.005% Dimethoate, 1.5 ml/L of Phosalone, 1.5 ml/L of Monocrotophos 36 WSC, 0.5 ml/L of Dichlorvos.

» Drip-irrigate with Dimethoate 30 EC @ 2 ml or Phosalone 35 EC @ 2 ml /L of water, then spread 10 t/ha of well-rotted sheep dung or 2 t/ha of poultry manure.

Rhizome fly, *Mimegralla coerrulelfrons*

Damage symptoms

» Maggots cause rhizome rot and plant yellowing by feeding on the plant's underground stems.

» Aerial plants have wilted and dried up, and their hearts have died as a result of primary harm.

» Rhizomes rot when the fungus *Pythium* sp. invades them via feeding injuries.

Identification of pest

» Tiny, white, cigar-shaped eggs that narrow at both ends.

» Larvae in maturity are a podous, creamy white colour and measure 9.5 mm in length and 1.95 mm in width.

» Adult flies tend to be somewhat sizable, with a long, lean body. The wings are see- through and spotted with an ashy grey, while the body is solid black. Fly wing span can be anywhere from 13 mm to 15 mm.

Favorable conditions

» Weather is warm and muggy.

Management

» Clearing off unwanted vegetation during the dormant months.

» The process of picking a robust rhizome for a new garden.

» After the crop has been harvested, the rotting rhizomes and any maggots they may contain must be removed from the field and destroyed.

» Using a spray containing either 0.05% Fenitrothion of Monocrotophos or 0.02% Diazinon.

» Beginning in July, spray Methyl Parathion at a rate of 0.05%, twice or thrice.

White grub, *Holotrichia* spp.

Damage symptoms

» The grub consumes the plant's roots and young rhizomes.

» Insect infection causes leaf yellowing.

» Pseudo stems may be severed at their bases if the situation warrants it.

» In heavily affected plantations, the entire crop can be wiped out.

Identification of pest

» Adults are roughly 2.5 mm × 1.5 mm in size, and their dark brown coloration is unmistakable.

» Creamy white grubs can be found in the ground.

Management

» Pest populations can be reduced by leaving land fallow for two years in a row.

» The increase in grub population can be slowed by growing sunflowers and other grub-resistant crops.

» Planting sorghum, maize, onions, and other "trap crops" to limit the spread of white grubs.

» Spreading *Beauveria bassiana* or *Metarrhizium anisopliae* in vermicompost at 5 g/kg or soaking the soil with these entomopathogenic fungi at 5 g/L.

» Neem oil 0.15 EC (1500 ppm) @ 3 ml/L was sprayed twice at 15-day intervals, and showed significant improvement.

» Adults can be lured into light traps and killed in this way.

» The grubs can be managed by combining fine cow dung with the entomophathogenic fungus *Metarrhizium anisopliae* and then applying the mixture on the field.

» Neem cake should be applied to the soil at a rate of 40 kg per hectare just before planting crops in endemic regions.

DISEASES

Rhizome rot, *Pythium aphanidermatum*

Symptoms

- » The infection spreads from the pseudo stem's collar area in both directions.
- » As the water seeps into the collar area of the damaged pseudo stem, soft rot begins to spread down into the rhizome.
- » Infected rhizomes, which eventually decay, are also observed later on.
- » Foliar symptoms manifest as a subtle yellowing of the leaf tips and leaf blades.
- » Early on in the disease's progression, the disease only affects the outside edges of the leaves, leaving the heart of the leaf unaffected.
- » All of the plant's leaves, beginning at the base, become yellow, and the pseudo stem eventually droops, withers, and dries out.

Management

- » Since the plant is more vulnerable to infection when water sits in one place for an extended period of time, cultural techniques such as planting in well-drained soils are crucial for disease control.
- » Since the disease can also be spread through seeds, it is important to source disease- free rhizomes from healthy gardens.
- » Leave as least four years between plantings of ginger on the same plot of land.
- » Grow corn, yams, and cassava that are resistant to soft rot.
- » It is possible to treat rhizomes for 60 minutes in a solution of 250 gm of Mancozeb (Dithane M-45) and 100 gm of Carbendazim (Bavistin) in 100 litres of water, before allowing them to dry in the shade for 48 hours.
- » Carbendazim at 0.2 percent applied to rhizomes resulted in the highest germination rate (91.33 percent), lowest disease incidence (10.66 percent), and highest yield (4.40 kg per pot).
- » Metalaxyl 0.25 percent, Mancozeb 0.3 percent, or copper oxychloride 0.3 percent drenched into the soil once before a monsoon is recommended.
- » The combination of solarizing the soil, dipping the rhizomes in 0.3% Metalaxyl, and applying *T. harzianum* and Eupatorium leaves to the soil yielded the best results.
- » *Trichoderma harzianum*, applied alongside Neem cake at the rate of 1 kg

per bed, has been shown to be effective against the fungus.

» Make use of cultivars that is tolerant of stress. In addition to SG 600 and Sadhupul and Jamaican Awacho and B 1, we now have SG 1 and British First Language B.

Fusarium yellows, *Fusarium oxysporum* f. sp. *zingiberi*

Symptoms

» Constantly occurring in clumps or waves.

» Dry rot, in contrast to rhizome rot, typically manifests in the field as localised, slow-spreading patches.

» The damaged plants are stunted and their leaves are fading to varied degrees.

» The older leaves tend to dry up first, followed by the younger ones.

» When the rhizome is at an advanced stage, a brownish ring can be seen around the cortical area when it is cut open.

» In contrast to soft rot, the pseudo stem of a plant with dry rot will not come off with a light tug.

» The disease causes the rhizomes to shrink and dry up, making them unsellable.

Favorable conditions

» Inadequate drainage and heavy precipitation are conducive to the spread of disease.

» Wet soil is ideal for farming.

Survival and spread

» Rhizomes harbouring fungi can be used as vectors.

» Furthermore, the fungus creates dormant structures (chlamydospores) in the rotting tissues of infected rhizomes.

» Therefore, the fungus is maintained in the field by tissues from infected crops that have not been harvested.

» Once the soil in a ginger field has been contaminated by a fungus, it can linger there for years.

Management

» Raised beds are advised.

- Preparing the soil for solarization months before planting is essential.
- Choose rhizomes that haven't been affected by any rot or other disease.
- Apply a 3% solution of Mancozeb or a 3% solution of Carbendazim on the seed rhizomes and water the soil thoroughly.
- When the problem arises again, treating the seeds with hot water (51 °C for 10 minutes) and then with Bordeaux mixture (1%), efficiently fixes it.
- Repair the drains.
- Mustard oil cake applied to the soil at a rate of 40 kg/ha prior to sowing in furrows has been shown to reduce nematode populations.
- Biological management employing helpful fungi (*Trichoderma* spp.) and bacterial isolates (*Bacillus* sp. and *Pseudomonas fluorescens*) (*Bacillus* sp. and *Pseudomonas fluorescens*).

Leaf spot / blight, *Phyllosticta zingiberi*

Symptoms

- Younger leaves often have a few tiny spindly to oval dots.
- Yellow rings encircle white, papery centres and dark brown edges in these locations.
- Spots eventually grow in size and merge to produce larger ones, reducing the available photosynthetic surface area. In the case of serious infection, the entire leaves dry out.

Favorable conditions

- Infectious agents in soil are a major cause of disease.
- From July to October, it can be seen on the leaves.
- Extremely hot and humid conditions.

Survival and spread

- Traveled via gusts of wind and drenching rain.

Management

- Choosing resistant seed rhizomes and applying either Carbendazim + Mancozeb or Carbendazim (0.25%) prior to sowing
- The plant population can be kept under tight control by cultivating it in somewhat shaded conditions.

- In the first 15 days of the disease, spray the Bordeaux mixture three to four times.
- Copper, Chlorothalonil, Tebucunazole, Mancozeb, Captan, and Tebucunazole spray.

Bacterial wilt, *Ralstonia solanacearum*

Symptoms

- At the collar region of the pseudo stem's base, the first dark black, water-soaked patches form and spread upwards and downwards.
- The lower leaves begin to show mild drooping and curling of the leaf margins, and this trend eventually rises to the upper leaves.
- The lower leaves are the first to turn yellow, and the top leaves follow.
- The disease then spreads to the leaves, turning them a bronze yellow tint and eventually causing the clumps to decay.
- When gently pushed, the injured pseudo stem and rhizome discharge a milky substance.

Management

- Choose planting areas with good drainage and use those soils.
- Choose rhizomes for planting from a disease-free field.
- Consider employing soil solarization, in which damp soil is covered with a transparent polythene sheet and left undisturbed for 45-50 days before planting.
- Non-host crops, such as rice, should be rotated in and out regularly.
- Don't utilise anything that has been tainted, including irrigation water and tools.
- Before being planted, ginger rhizomes are solarized for two to four hours (during the summer, this raises the rhizome temperature to 55°C).
- It is recommended that Mancozeb (3g/kg) be applied to the seed rhizomes before they are stored.
- Remove the infected clusters and spray the affected areas and nearby plants with Copper oxychloride 0.2%.

References

Dohroo, N. P. (2005). Diseases of ginger (pp. 305-340). Boca Raton: CRC Press.

Meenu, G., & Jebasingh, T. (2020). Diseases of ginger. Ginger Cultivation and Its Antimicrobial and Pharmacological Potentials, 1-31.

Ravindran, P. N., & Babu, K. N. (Eds.). (2016). Ginger: the genus Zingiber. CRC press.

Gopi, R., Avasthe, R. K., Kalita, H., Kapoor, C., Yadav, A., Babu, S., & Das, S. K. (2016). Traditional pest and disease management practices in Sikkim Himalayan region. International journal of bio-resource and stress management, 7(Jun, 3), 471-476.

17

Mustard

INSECT PESTS

Aphid, *Lipaphis erysimi*

Damage symptoms

» When they feed, they can completely cover the surface of flower buds, shoots, pods, etc.

» Both nymphs and adults feed on plant sap.

» The effects of overpopulation on plants can be devastating.

Pest identification

» These insects are tiny, soft-bodied, and grey-green in colour; their bodies are covered in a white waxy covering.

Favorable conditions

» Temperatures between 4 and 13.8 °C, highs of 14 to 30 °C, and humidity levels between 48 and 80 percent are ideal for rapid aphid reproduction.

Management

» If you want to lessen the impact of aphids on your crop

» Removing aphid-infested branches early on at the community level can help stop the problem from spreading further.

» Aphids can be discouraged from feeding on plants by using reflective mulches, such as silver coloured plastic.

- » The most effective way of control is the use of insecticidal soaps or oils, such as Neem or Canola oil.
- » When insect populations reach ETL, a spray of 0.03% Dimethoate or 0.025% Oxydemeton Methyl is applied.
- » There are many predators that prey on the mustard aphid, including as the syrphid fly, syrphid wasp, and *Chrysoperla carnea.*

Saw fly, *Athalia lugens proxima*

Damage symptoms

- » The larva feeds by nibbling at the edges of the leaf and eventually moves in towards the midrib.
- » Because of their ravenous appetites, the grubs eat huge holes in the leaves and often the entire surface.
- » They feed on the outer layer of the shoot, causing the plant to wither and die.
- » Drops in yield might be as high as 18%. Crop replanting is necessary if the seedling stage is severely affected.

Pest identification

- » Larvae have a wrinkly, splotchy, dark green to black body and eight pairs of prolegs. The larva immediately collapses to the floor and plays dead at the slightest provocation.
- » The wings of adult moths are a smoky transparent colour, while their bodies are orange.

Management

- » Summertime ploughing to eradicate the larval stage.
- » It's best to start planting right away.
- » Keep your farm tidy.
- » Sawfly control with irrigation is most effective when applied at the seedling stage, when most larvae are killed by the drowning effect.
- » Saw fly larvae are collected and killed twice daily.
- » To protect sawfly larvae, it is important to keep *Perilissus cingulator* (a parasitoid) and *Serratia marcescens* bacteria under check.
- » An anti-feedant emulsion made from bitter gourd seed oil.

» Imidacloprid 70% WS @ 700 g/100 kg seeds Malathion 50% EC @ 600 ml in 200- 400 L of water/acre Methylparathion 2% DP @ 6000 g/acre Dimethoate 30% EC @ 264 ml in 200-400 L of water/acre Quinalphos 2 5% EC @ 480 m in 200-400 L of water/acre

Painted bug, *Bagrada hilaris*

Damage symptoms

» It is a pest that shows up occasionally during the seedling and harvesting phases of cruciferous crops.

» In the face of a pest onslaught, young plants quickly droop and die.

» The adult insects produce a sticky secretion that rots the pods.

» When severe incidence occurs during the seedling stage, it may be necessary to resow the crop during the harvesting phase.

» The oil and seed mass are both diminished as a result.

Pest identification

» Nymphs go through five distinct transformational stages, beginning as a bright orange and ending as a deep red with dark markings. They are born without wings but develop them as they mature into nymphs. Nymphs in their final instar stage can be identified by the appearance of wing pads.

» The normal dimensions of an adult shield insect are 5–7 mm in length and 3–4 mm in width. Black, white, and orange marks are spread across the surface.

Favorable conditions

» Between 21 and 40 °C, it maintains its functionality.

» February and March rains helped it flourish.

Management

» The eggs of painted bugs must be killed, thus deep ploughing is necessary.

» In order to protect crops from pests, they must be sown early.

» Weeding out old crop debris from the field is one way to reduce the prevalence of the painted bug.

» Weeding on a regular basis helps keep painted insect populations from skyrocketing.

» Reduce pest damage by watering the crop four weeks after planting.

- Bugs tend to concentrate on the plant's leaves and stem, so if you jerk those parts, the bugs will fall off and can be destroyed in kerosene water.
- To eliminate painted bugs, spray either 0.05% Malathion or Dichlorvos76% EC at 250.8 ml in 200-400 L of water/acre, 700 g of Imidacloprid per 100 kilogm of seed, or 6 kilogm of Phorate10G per acre.
- Accelerated threshing of the crop after harvest.

Diamondback moth, *Plutella xylostella*

Damage symptoms

- White spots on leaves are caused by immature larvae scraping the epidermis.
- In later stages, larvae burrow holes in the leaves, giving them a shrivelled appearance.
- The leaves could be totally consumed.
- It can also be used to feed maturing seed by boring into pods.

Pest identification

- Larvae are a yellowish green colour and are covered in fine, upright black hairs.
- The adults of this species are tiny and a drab greyish brown, with slender, white wings that are edged in yellow. The inner border of the forewings is marked by three white triangle dots. The diamond-shaped markings on the triangular wings are really found on the opposing wings. Long, delicate hairs fringe the back wings.

Management

- Put in pheromone traps at the rate of five per acre to keep an eye on the moth population.
- At least twice a week, you should collect the larvae when they are at the gregarious stage and then carefully destroy them.
- The diamondback moth parasitoid *Cotesia plutellae* should be protected. The diamondback moth's most significant parasitoid is the nematode *Diadegma insulare*.
- Apply 5% Malathion dust at 37.5 kg/ha to remove adult larvae.

Leaf webber, *Crocidolomia binotalis*

Damage symptoms

- Larvae feed on the chlorophyll in immature leaves, then move on to older leaves, buds, and pods where they spin webs and spend their time.
- Defoliation occurs on severely damaged plants.
- Insects and rodents nibble away at seed pods, destroying their contents.

Pest identification

- Larvae are a light yellowish brown colour with a scattering of small black patches over their sides and undersides.
- The adult moths have a yellowish brown colour, with reddish brown, distinct and indistinct wavy lines on the forewings, and large white spots towards the thorax that are surrounded by a black tuft of hair. The tips of the white hind wings are a dark brown colour.

Management

- Fenvalerate 20 EC 0.5 ml/L or Phosalone 35 EC @ 2 ml/L can be sprayed.

Head borer, *Hellula undalis*

Damage symptoms

- The leaves are first mined by caterpillars, turning them into a papery white colour.
- Soon after, they begin boring into stems and feeding on leaves.
- Silk and faeces cover the opening.

Pest identification

- Larvae are a light brown colour with four or five longitudinal stripes that are pinkish brown.
- The adult moths of this species are a very light greyish brown with a rosy undertone. Grey undulating lines can be seen on the forewings, along with a spot at the very tip and a lunule with a pale centre and a dark outside edge. The tips of the back wings are darker than the rest of the feathers.

Management

- Twice weekly leaf collection and killing of gregarious-stage larvae.

- Apply 5% Malathion dust at 37.5 kg/ha or Carbaryl 50 WP at 375 g in 150 litres of water to kill off mature larvae.

DISEASES

White rust, *Albugo candida*

Symptoms

- Pustules ranging in colour from white to a creamy yellow appear on the undersides of leaves.
- The pustules eventually merge into larger areas on the underside of the leaf.
- On the upper surface of the leaf, in the opposite direction of the pustules, tan yellow patches begin to appear.
- Also, pods are prone to developing pustules.
- Flowers that have been infected with this fungus may develop a deformed, green appearance known as a "stag head," and may have thick, club-shaped, and larger versions of their normal floral components.

Favorable conditions

- Field circumstances most conducive to the spread of the disease include temperatures around 12 °C, relative humidities around 70%, wind speeds between 2.7 and 3.4 km/h, and precipitation throughout the winter months.

Management

- To reduce the spread of disease, it is best to clear the field of any remnants of the previous crop.
- Reduce disease spread by planting your crop early.
- Rotate the crops every three years at the very least.
- Weed out the susceptible mustard-type plants and the volunteer canola in your crop rotation.
- *Trichoderma viride* seed treatment at 4 g/kg.
- Taking out of the field any plants that are severely affected with a disease.
- The bacterium *T. viride* was applied to the soil at a rate of 62.5 kg of FYM per hectare (kg/ha).
- A 0.2% Mancozeb spray can be used as needed.

Downy mildew, *Peronospora parasitica*

Symptoms

- Necrotic (dead) dots on leaves start out looking angular and translucent pale green before turning greyish white.
- The flower cluster stems swell up.

Management

- Eliminate farm waste.
- Replacement of host plants by crop rotation.

Powdery mildew *Erysiphe cruciferum*

Symptoms

- Small white dots emerge on leaves, and then they spread until the entire leaf is covered in white powder.
- Although the disease initially manifests itself on the underside of the plant's elder leaves, it eventually spreads to all of them.
- Defoliation occurs early due to the withering and death of infected leaves.
- As a result, yield is diminished because less light can reach the plant's leaves.

Management

- Apply 20 kg/ha of 300-mesh Sulfur at the first sign of diseased, and then again 15 days later.
- Spraying with 0.2% wettable Sulfur (Sulfex 25 g/10 L) or 0.25% Dinocap (Karathane 5 ml/10 L) at the first sign of disease, then again 15 days later, is an effective and cost-efficient method of disease control.
- In the morning, spray a mixture of mineral oil and the AQ 10 biofungicide (*Ampelomyces quisqualis*).

Leaf spot, *Alternaria brassicae, A. raphani*

Symptoms

- Grey dots on leaves, encircled by a yellow halo, are the first sign of a disease.
- Leaf and pods grow these black, round markings a few weeks before they're ready to harvest.

- Possible contamination of seed stock.
- It's possible for there to be dark areas within a ring-like pattern of focus.
- Pods break easily because of the severe black spot diseases they often suffer.

Management

- While crop rotation can lessen the spread of disease from one field to another, it cannot completely prevent the spread of disease through the air.
- Reduce the viability of shrunken seed, which may be infected with black spot, by thoroughly cleaning the seed.
- Make sure the seed you're using has a germination rate of over 90%.
- Damage from black spot can be reduced by swatting infested crops as soon as possible.

Stem rot, *Sclerotinia sclerotiorum*

Symptoms

- Sometimes a cottony white growth covers the water-soaked patches on the stems.
- The condition causes the damaged stem tissue to bleach and eventually deteriorate.
- When viewed from a distance, the internodes and the base of the plant seem white.
- Stem shrivelling, wilting, and drying out before its time, and premature ripening.
- In the later stage, the diseased plant portions may also show sclerotial bodies, which range in colour from brown to black.
- Premature ripening and lodging are the results of stem girdling.

Favorable conditions

- The conditions for the spread of the disease are ideal: constant wind, an average temperature range of 18-25 °C, and a high humidity range of 90-95%.

Survival and spread

- The pathogen can persist as mycelium in both living and dead plants, as well as sclerotia in infected plant parts, on the soil's surface, or in contaminated seed.

Management

- Cereals should be rotated with other crops; extremely sensitive crops such dry edible beans, sunflowers, mustard, and canola should not be planted more than once every four years. Rotate heavily infested fields every several years for at least five years.
- Planting near an area that has suffered from Sclerotinia heavily within the past four to five years is discouraged.
- Planting in evenly spaced rows will help prevent a thicket of vegetation.
- Remove the broad-leaf weeds.
- Incorporate freshly sterilised seed into the ground.

References

Yadav, M. S., Godika, S., Yadava, D. K., Ahmad, N., Mehta, N., Bhatnagar, K., ... & Chattopadhyay, C. (2019). Prioritizing components of package of integrated pest management in Indian mustard (Brassica juncea) in India for better economic benefit. Crop Protection, 120, 21-29.

Kolte, S. J. (2009). Management of major diseases and pests of mustard in India. Bulletin: Consultative Group for Research on Rapeseed, 25.

Agrawal, R., & Mehta, S. C. (2007). Weather based forecasting of crop yields, pests and diseases-IASRI models. J. Ind. Soc. Agril. Statist, 61(2), 255-263.

Gautam, M. P., Singh, S. N., Kumar, P., Yadav, S. K., Singh, D. P., & Pandey, M. K. (2019). Mustard aphid, Lipaphis erysimi (Kalt)(Hemiptera: Aphididae): A review. The Pharma Innovation Journal, 8(9), 90-95.

Thomas, J., Kuruvilla, K. M., & Hrideek, T. K. (2012). Mustard. In Handbook of herbs and spices (pp. 388-398). Woodhead Publishing.

Yadav, S., & Rathee, M. (2020). Sucking pests of rapeseed-mustard. Sucking Pests of Crops, 187-232.

Kumar, A., Agrawal, R., & Chattopadhyay, C. (2013). Weather based forecast models for diseases in mustard crop. Mausam, 64(4), 663-670.

18 Rubber

INSECT PESTS

Bark feeding caterpillars, *Aetherastis circulata, Ptochoryctis rosaria*

Damage symptoms

» The caterpillar hides in a web and eats the rotting bark from the trunk.

» When it eats the bark that is supposed to be rejuvenating, it becomes a problem.

» Throughout the tree's trunk and branches, caterpillars construct silken and feces- lined passageways.

» Larvae can cause latex exudation from either dead or live bark when they eat it.

» Areas of deep scarring are observed at feeding sites.

Pest identification

» The caterpillars are just 1.5 to 2 cm in length and are a brick red colour with a flattened shape, a large head and thorax, and a narrowing body.

» Adults of this species are a little white moth with a few dark spots on each forewing.

Management

» Dispose of larvae and mud tunnels simultaneously.

» Carbaryl 100 g and 1 L water. Swipe the affected area with a paintbrush dipped in this solution.

- If the infestation is severe, use a power duster to spread Carbaryl 5% dust at 10 kg/ ha or Fenval 0.4% dust at 7 kg/ha.
- Apply 0.02% Fenvalerate spray to the tree trunk.

Cockchafer grub, *Holotrichia serrata*

Damage symptoms

- The cockchafer grub eats the roots of seedlings and young plants, causing them to die.
- Infestations of grubs typically cause damage to the tap root.
- Unaffected seedlings' leaves become yellow and fall off, whereas those that have been infected do not.
- Everything about the plant dries up and dies off later.

Pest identification

- Grubs are about the size of a dime, have a broad 'C' shape, and are a drab white colour.
- The adult form of a beetle can be any of several metallic hues, including blue, green, greenish brown, or brown.

Management

- Adult beetles can be captured using specially built light traps.
- During the time you are preparing your nursery beds, mix 25 kg of Phorate 10G per hectare into the soil.
- Soak the root zones of afflected trees with 2 g/L of Chlorpyriphos 25 EC.
- Entomopathogenic fungus *Beauvaria brongniarthie* and bacteria *Bacillus popillae* are used for biological control of grubs.

Mealy bug, *Ferrisia virgata*

Damage symptoms

- Colonize the rubber tree's leaf undersides.
- Affect a young, vulnerable shoot, draining it of its life force.
- The first symptoms of a mealy bug infestation are leaves that look wilted or dried out and cottony masses at the leaf attachment sites and the undersides of the leaves.

- Similar to the occurrence and damage caused by scale insects.
- The presence of sooty mould is also linked to this phenomenon.

Pest identification

- Small insects with a mealy, white exterior are known as mealy bugs.

Management

- Treating for mealybugs requires manual extermination.
- Apply a 25 g/L solution of fish oil rosin soap.
- There should be no more than 2 ml of Monocrotophos per L of water when using a sprayer.
- For similar results, you can use organophosphorus insecticides like Malathion 0.1% (2 ml/L) or Quinalphos 0.075% (Ekalux 25EC at 3 ml/L) as a spray.

Scale insect, *Saissetia nigra*

Damage symptoms

- The outer black covering of scale insects is fashioned like a dome.
- Sap the fragile twigs, petioles, leaflets, and shoot parts, causing them to dry out.
- Infest the branches and leaves, resulting in leaf drop.
- Those areas that are severely impacted dry out and die.
- This bug is linked to black mould and ants.

Management

- This pest is kept in check by natural enemies like insect parasites and entomopathogenic fungi.
- Spray a 0.05% concentration of an organophosphorus insecticide, such as Malathion, when a severe infestation is discovered.

Termite (white ant), *Odontotermes obesus*

Damage symptoms

- Lives off of the decaying bark of trees and the new shoots of plants.
- Constructs a network of earth tunnels with a cover over the tapping panel and the collection cup.

- Sometimes when young plants are attacked they dry out.

Management

- Chlorpyriphos 0.1% solution should be drenched into the soil around the damaged plants' root systems.

Slugs and snails, *Mariaellae dussumieri, Crytozona bistrialis*

Damage symptoms

- Young plant growth is halted when they ascend the stems and feast on the terminal and auxiliary buds.
- The resulting cluster of shoots has the shape of a club.
- They occasionally use the tapping cut and collection cups to climb older trees and drink latex.
- During the day, they hide out beneath mulch or other debris, waiting for nightfall to come so that they can feed.
- Spring-like faeces and shimmering lines of dried sticky secretion are telltale signs of their presence.

Management

- Spray the bottom of the stem with metaldehyde paint or scatter bait pellets containing 2.5% metaldehyde (Snailkill 2.5) over affected plants and seedling beds.
- Brushing with 10% Bordeaux can deter slugs and snails. Cut a 30-centimeter-long section of paste and wrap it around the stem just above the bud union. If the attack persists after 30–45 days, apply again.
- Metaldehyde is more effective when sprayed directly on an affected region rather than distributed.

DISEASES

Abnormal leaf fall, *Phytophthora palmivora*

Losses

- This disease can reduce rubber crop yields by 9-16% in susceptible clones.

Symptoms

- While most of the leaves drop in December, this phenomenon is most noticeable between June and August.
- Discolored grey patches develop irregularly on leaf surfaces.
- Rotting occurs in affected fruit.
- The affected leaves fall to the ground in a dense carpet of rotting vegetation that gives off a foul odour.
- The fruits rot first, then a lot of infected leaves fall off before they're supposed to, either while still green or after they've become a coppery red colour.
- Sometimes a black lesion will form on the petiole (showing as a depressed spot) and a drop of latex will coagulate in the centre.
- The lesion may spread to the leaf's midrib and blades.
- Extreme defoliation can cause terminal twig death and significant crop loss.

Favorable conditions

- The most significant predisposing meteorological component is precipitation. It's peak visibility months are June, July, and August.

Management

- Spraying foliage with 1% Bordeaux mixture (the addition of $ZnSo_4$ @ 0.2% increases its effectiveness) using high volume sprayers or with oil-based Copper oxychloride dispersed in diluents spray oil using low volume air blast sprayers from the ground or from the air is a preventative measure taken prior to the onset of the southwest monsoon.
- Tolerant responses to disease have been observed in rubber clones RRII 105, PB 217, GT 1, and GL 1.

Birds' eye leaf spot, *Drechslera heveae*

Symptoms

- Small, necrotic, circular patches with a grey centre and a dark/brown border appear frequently on young plants in the nursery or in the field.
- An infection of this magnitude causes the plant to defoliate and die back prematurely.
- Matured seedlings in nurseries are at risk.

Favorable conditions

- When conditions are dry, this disease thrives.
- Diseases thrive in the sun.
- The risk is greatest for plants growing in low-fertility soil.

Management

- Nurseries with overhead shading have lower rates of plant disease.
- Strong seedling health can be preserved with careful, well-balanced feeding.
- Spraying with a 1% Bordeaux combination, 0.2% Mancozeb (Mancozeb/Indofil M-45 at 2.5 g/L), or 0.02% Carbendazim (Baiting at 0.4 ml/L) multiple times.

Powdery mildew, *Oidium heveae*

Symptoms

- Both immature and fully grown leaves develop a white, powdery fungal growth.
- Curling, wrinkling, and rolling inward before falling off, infected leaves give the tree a broomstick appearance.
- Flowers and young fruits are particularly susceptible, as they drop their petals and ripen their fruit when infected.
- The disease is more common in plants that are grown in the shade.
- Extreme and persistent infection causes branch die-back and ultimately tree mortality.

Favorable conditions

- Disease is more likely to spread between 25 and 30 °C.

Management

- The recommended frequency of sulphur dusting or spraying is every 15 days.
- Spraying with either 0.1% Carbendazim or 1.5% Tridemorph and then dusting with 70% Sulfur.
- The combination of Carbendazim, wettable sulphur, and microsul at a concentration of 52% EC as a spray is similarly effective.
- Among the cultivated clones, PB 86, GT 1, PR 107, RRII 208, and PB 310 are the least vulnerable.

Leaf spot, *Corynespora cassiicola*

Symptoms

- Outwardly, it looks like there are big brown spots with a white centre.
- It's possible the centre will collapse, creating shooting lanes.
- Thin, new, light green leaves that appear on established trees during refoliation are especially vulnerable.
- A cluster of tumours has grown into a huge, discoloured area.
- As the disease progresses along the veins, a brownish "railway track" or "fish bone" appearance develops.
- Abscission of leaves can occur from even a localised infection on the midrib or base of a leaf.
- When leaves fall, branches die back.

Favorable conditions

- When combined with warm daytime temperatures (34 to 36 °C), excessive humidity is ideal for the spread of the disease.

Management

- In repeated spraying, either 1% Bordeaux mixture, 0.2% Mancozeb, or 0.02% The nursery should use carbendazim.
- A shady nursery has a lower disease rate.
- In the field, spray a 1% Bordeaux mixture.
- Mancozeb 0.2% (Dithane/Indofil M-45 at 2.66 g/L), Copper oxychloride 0.2%, or Carbendazim 0.05% (Bavistin at 1g/L) sprayed at high volume every two to three weeks during refoliation in a mature plantation is highly effective.

Colletotrichum leaf disease, *Colletotrichum acutatum*

Symptoms

- Diseased leaves get misshapen, blacken, shrivel, and fall off the plant, although the petioles remain attached for a while.
- Most leaf infections begin at the leaf tips and work their way down the leaves.
- Usually, the necrotic part of the leaf is the tip and the margins, and the rest of the leaf just gets wavy.

- A little brown spot with a brown edge and a yellow halo appears on the leaves that have been affected.
- At the mature stage of the leaf's development, the spots become thick and elevated, taking on a conical shape.

Favorable conditions

- The fungus can only produce spores in conditions of high humidity.
- Wet conditions are ideal for the spread of the disease. When fresh leaves are developing, heavy rain is frequently the cause of a severe attack.

Spread

- Splashes of rain disperse and transport the spores.
- Wind, fog, and precipitation all aid in the propagation of spores.

Management

- Upkeep of the field includes regular weeding, especially around the (disc) plants, and cleaning of any debris.
- A lack of ponding water in the garden indicates poor drainage.
- You can use a spray containing 0.2 percent Mancozeb, Carbendazim (Carbendazim 500 FL at 3%) or Hexaconazole (Saizole 5 SC at 2-3%) or Saipora SC 350 (mixed Carbendazim 300 g/L v 50 g/L), or a mixture of these at 0.2 to 0.5%. When the disease is first spotted, spray the young leaves, and continue spraying cycles every 7-10 days, especially after heavy rains.
- At 10- to 15-day intervals, spraying with 1% Bordeaux mixture, 0.125% Copper oxychloride (Fytolan 2.5 g/L), 0.20% Mancozeb (Dithane/ Indofil M-45 2.66 g/L), or 0.05% Carbendazim (Bavistin 1 g/L) is successful. Every two weeks, do another round of spraying.
- Field tolerance can be shown in the rubber clones PB 217, PB 260, and RRIM 600.

Pink disease, *Corticium salmonicolor*

Symptoms

- The bark may develop a white or pinkish cobweb mycelial growth with latex streaks pouring out of wounds.
- The affected bark eventually rots, dries out, and cracks.

» New growth appears beneath the damaged area.

» Dried, dead leaves cling to the remaining twigs and branches at their tips.

Favorable conditions

» High relative humidity and wet conditions are ideal for the spread of disease.

Management

» In order to lessen the spread of the disease, field cleanliness is essential.

» The inoculum can be reduced by pruning and eliminating damaged branches.

» Fork and branch spraying with a 1% Bordeaux mixture twice a year (May and August) significantly reduces disease spread.

» After applying Bordeaux paste, the damaged area can be scraped to remove mycelium and rotting bark, and then a second application of Bordeaux paste can be made.

» Wound dressings can include Thiram 0.75 percent [Thiride (TMTD) at 10 g/L] in 1% ammoniated field latex or Tridemorph 2 percent (Calixin at 25 ml/L) in 1% ammoniated field latex or Thiride 0.75 percent or Tridemorph 1 percent or Propiconazole 0.1 percent (Tilt at 4 ml/L) in Pidyvyl, china clay, and water (1: 2: 4 by volume).

» Clones of rubber with higher resistance include PB 86, RRIM 513, GL 1, PR 107, GT 1, and PB 260.

Patch canker or bark canker, *Phytophthora palmivora, P. meadii,* and *Pythium vexans*

Symptoms

» The bark swelled and burst, releasing a sticky, amber latex liquid.

» Between the wood and the rotting bark, there is a coagulated rubber pad that gives out an unpleasant odour.

» Once this is taken off, you may also notice a little change in the wood's colour here.

Management

» Mancozeb 0.75% (Dithane/Indofil M-45 at 10 g/L) can be applied to the wound after the affected area has been scraped to remove all rotten bark and coagulated rubber.

» Dress the wound with a compound dressing after the fungicide has dried.

Brown root disease, *Fomes noxius*

Symptoms

» A thick layer of soil and pebbles, kept in place by a mycelium network, has accumulated on the roots of the infected plants.

» All of the tree eventually dies from the rot that spreads to the tap root.

Management

» Cut down and destroy the diseased trees.

» Distribute lime at a rate of 2.5 t/ha

» Apply an extra dose of lime in the vicinity of the dead tree, at a rate of 25 kg per tree.

» Use 0.1% Emisan or Aretan to clean the affected area of the root.

References

Wastie, R. L. (1975). Diseases of rubber and their control. PANS Pest Articles & News Summaries, 21(3), 268-288.

Rao, B. S. (1974). Diseases and pests of rubber in South and South East Asia. *World Crops*, *26*(2), 75-81.

Chee, K. H. (1990). Present status of rubber diseases and their control. Review of plant Pathology, 69(7), 423-430.

Bastos, J. A. M., Flechtman, C. H. W., & de Figueiredo, R. W. (1979). Contribution to knowledge of the pests of Ceara rubber. Fitossanidade, 3(1/2), 45-46.

19 Sweet Potato

INSECT PESTS

Weevil, *Cylas* spp.

Damage symptoms

» Vegetation wilting and larvae eating tubers are two symptoms of this problem.

» Cavities, a spongy texture, and a darkening of colour develop in the infected tuber.

» Even minimal feeding causes a chemical response that leaves the tubers tasting bitter and smelling of terpenes.

» To make matters worse, larvae mine the tuber, causing it to discolour, break, or even collapse.

» In order to access the tubers, the adult may bore countless tiny holes, each around the size of its head.

» Reduced yield, shorter shelf life, and diminished plant vitality are all possible outcomes.

Pest identification

» The whitish larvae have no legs and are a peculiar shape.

» The adult insect has a body length of 5.5-8.0 mm, a black head, orange or reddish brown antennae, thorax, and legs, and a metallic blue abdomen and elytra. The length of the snout is roughly equal to that of the chest, and it is somewhat curved.

Management

- Make sure to plant with materials that haven't been infested by weevils.
- Take out and destroy the ruined crops.
- To prevent the weevil from establishing itself, incorporate planting mounds or ridges that have been covered with a thick layer of fresh Lantana camera leaves.
- At high relative humidity and insect densities, the fungal disease *Beauveria bassiana* has been shown to cause substantial mortality rates.
- When applied at a rate of 3 t/ha, mulches made of plastic, rice straw, Eupatorium or Clerodendron leaves keep the weevils from reaching the roots through crevices in the soil.
- Planting cycles that include rice, sorghum, yam, and edible sunflower.
- Sweet potatoes and marigolds (a barrier crop), garlic, and coriander were planted one to one, whereas yam beans and lima beans were planted two to one.
- Field traps containing a pheromone-infused lubricant and tiny rubber containers.
- It is advised to apply pesticides to the soil, namely Carbofuran (1.4 kg/ha), 0.3% Diazinon, 1.5 kg/ha Disulfoton, 0.1% Fenthion, and 1.4-2.0 kg/ha Phorate.
- An application of 0.05% dimethoate on the foliage.
- *Heterorhabditis bacteriophora*, a species of entomopathogenic nematode, is a powerful tool for biological pest management.

Hornworm, *Agrius convolvuli*

Damage symptoms

- Caterpillars eat holes in leaves in a non-uniform pattern as they munch.
- It is possible for them to consume an entire leaf, leaving only the petiole behind.
- Frass left by insects can be seen in the soil close to the affected plant section.
- A single huge caterpillar is all it takes to completely strip a plant of its leaves.
- Caterpillars of a certain age can quickly strip a field of its foliage if there is a significant population of them.

» If severe defoliation is performed on a young crop, it can reduce yields.

Pest identification

» Eggs that are 1 mm in diameter and a pale green colour.

» Initially green with white markings, the caterpillar grows to a dark olive or brown with varying patterns. At full size, up to 120 mm in length.

» A huge (about 60 mm) adult moth, characterised by its mottled brown and grey wings. Unique abdominal stripes of black, pink, and white.

Management

» By exposing the pupae to predators and the sun's heat, turning the soil between crops is a bad idea.

» Put to good use Neem, Derris, Pyrethrum, and Chili, all of which come from plants (with the addition of soap). The first sign of larvae requires treatment with a 0.2% Carbaryl solution in a high volume spray, 0.05% Endosulfan, or Trichlorfon at 100 g a.i. in 100 L of water.

» Four species, *Trichogramma australicum, T. achaea, T. agriae,* and *Telenomus* sp., parasitize the eggs of *Aedes convolvuli. Trichogramma* spp., tiny wasps, are among the most important egg parasitoids; *Sycanus* sp., a huge reduviid (predatory bug), and a tachinid fly are among the larval predators.

» Spinosad (Success) and *Bacillus thuringiensis* var. *kurstaki* are two examples of synthetic products that contain disease-causing organisms that can be used to control pests and disease.

Armyworm, *Spodoptera litura*

Damage symptoms

» Larvae can consume entire leaves.

» They eat by gnawing on leaves until just the midrib remains.

» They start off social but develop an introverted streak.

» Initially, only the leaf's upper epidermis is chewed, but eventually, the entire leaf is eaten, except for the midrib, which is left untouched.

» Many of these caterpillars travel in large groups.

» Caterpillars consume the leaves of your crop, which could lead to a decrease in yield, both in terms of the number of tubers or leaves per plant and the size/ weight of those tubers or leaves, if a particularly bad infestation hits early on.

» The sweet potato weevil may be more active if the harvest is delayed due to leaf damage.

Pest identification

» The larva has 10 fleshy legs at the back and six smaller ones directly behind the skull. The lengths range from about 2.0 cm to about 4.5 cm.

» This moth is around 1.5 inches in length and its wingspan is 3 to 4 centimetres. The armyworm moth can be recognised by the little white dot in the middle of each forewing.

Management

» By reducing the availability of females for egg production, intercropping sweet potatoes with onion or the silver leaf (*Desmodium uncinatum*) could be beneficial.

» Keep a close eye on the harvest and pick out the bug eggs one by one.

» As soon as you notice gregarious caterpillars, spray them with a contact insecticide.

» Insect predators such as ants, ladybird beetles, lacewings, and dragonflies all prey on caterpillars.

» Caterpillars have been found with the fungus *Beauveria bassiana* on them in the field after a shower.

» During the early larval stage, when the larvae are still social, spot applications of pesticide or *Bacillus thuringiensis* can be made.

» The concept of HaNPV can be formulated in various ways.

» *Nomuraea rileyi*, often known as the green muscardine fungus, causes severe disease in *S. litura*.

Leaf miners, *Bedellia* spp.

Damage symptoms

» While the upper and lower membranes (epidermis) remain unharmed, the transparent larvae's diet consists of the leaf's green internal tissue.

» *Serpentine mines* (thin grey or silvery tracks) are made by young larvae, while "blotch mines" are made by older, more developed larvae.

» Once the mined tissues are eliminated, the serpentine or blotched mine will turn into a hole.

- Infested leaves will develop a grimy, black frass on the underside and exhibit silky webbings where the insect pupates.
- High levels of infection cause leaf browning.
- After a dry spell, or sometimes even during one, there may be a massive infestation.
- A decrease in effective leaf surface and hence in tuber yields can result from a severe breakdown.

Pest identification

- Larvae are easily identifiable because to their segmented body and rather pointed head and abdomen.
- Larvae can range in hue from crimson to a yellowish grey. Silken filaments ensnare the pupa.
- The adults of *Bedellia* spp. are tiny moths that range in colour from tan to grey and measure just 3 mm in length and 8 mm in wingspan.

Management

- Although *Bedellia* spp. are often managed by predators, foliar applications of insecticide may be effective in eliminating these plants.
- It has been suggested that the pest be managed using Carbaryl at 1 kg a.i. /ha, Chlorfenvinphos at 0.6-0.9 kg/ha, Diazinon at 55 ml/ha, Dimethoate at 14 ml/ha, Endosulfan at 70 ml/ha, Malathion at 14 g/ha, or Trichlorfon at 100 g/ha.

White grub, *Phyllophaga ephilida*

Damage symptoms

- The primary plant roots are a favourite food for white grubs.
- They go down into the soil and sever the plants' main stems and roots.
- The sweet potato's outer skin is a tasty meal for the larvae. The majority of harm occurs at this stage.
- They burrow inside of tubers.
- Tubers are damaged by galleries made by the larvae.
- These pests' diets weaken and ultimately kill plants by attacking their subterranean stems and roots.

- Unfortunately, the tubers are now worthless due to the tunnels that have formed in them.
- The tubers themselves are of lower quality.

Pest identification

- White grubs are characterised by their white or yellowish bodies, tan or brownish heads, and six spiky legs.

Management

- By ploughing in the late spring or early fall, you can eliminate the risk of the pests' eggs, larvae, and pupae surviving the process.
- Take care to guarantee enough drainage.
- Always rotate your crops. White grubs don't seem to bother soybeans very much.
- Insecticides such as Actara and Pronto have been used successfully to bring the problem under control.
- Grubs and adult beetles can be eliminated with the help of microbial insecticides like *Bacillus popilliae* and *Bacillus lentimorbus. B. popilliae* commercial preparations are highly specific for their host and have a long half-life.
- The parasitic wasp, *Pelecinus polyturator*, is one of the white grub's natural enemies.
- The grubs also become infected with Cordyceps fungus.

Striped weevil, *Alcidodes dentipes*

Damage symptoms

- Stems and tubers are both targets of the striped weevil's destructive appetite.
- The adults girdle the stems, which eventually causes the plants to wither.
- Galls grow on the stems as larvae dig into them, just like they do with Cylas larvae, although the tunnels they create are significantly bigger.

Pest identification

- The elytra of an adult *Alcidodes dentipes* are brown with four white streaks.

Management

- Rotating your crops is a smart farming practise.
- Plants should only be planted in sterile soil.
- Draining and cleaning the field.
- Picking at at the right time.
- A common method for controlling this pest is to regularly soil up the area around the plant's trunk and roots.
- Techniques described for Cylas should be effective against *Alcidodes* spp.

Bihar hairy caterpillar, *Diacrisia obliqua*

Damage symptoms

- Numerous larvae with yellow hairs skeletonize the leaves.
- Vegetables can be defoliated entirely.

Pest identification

- Larvae reach maturity at 30–40 mm in length and feature 7 orange bands across the body, each with tufts of yellow hairs.
- Cocoons with a rusty brown colour.

Management

- For effective pest control, use Endosulfan or Fenitrothion at a 0.05% concentration.
- To kill larvae, spray a 0.2% solution of diazinon, dichlorvos, or carbaryl.

Flea beetle, *Chaetocnema confinis*

Damage symptoms

- Young plants are a favourite snack for the flea beetles.
- Commonly found on the undersides of leaves, their feeding causes countless tiny holes of all shapes and sizes.
- Due to its small size and high activity levels, the beetle rarely stays in one place to feed for long.
- The young larvae eat roots.
- The tap roots have their hairs trimmed and round indentations made in them.

- Damage from its feeding is visible on the surface of the root, which might lower the value of tubers.
- The adults consume the young plants' leaves, stalks, pods, and blossoms.
- Their unique "shot hole" appearance in the leaves is the result of them chewing small holes or pits, often less than 3 mm in diameter.

Pest identification

- The larva is primarily white, but it has a brown head and three sets of brown legs close to its body.
- Its expanded black hind legs and slightly hairy wing covers distinguish the adult form. It measures roughly 2 mm in length.

Management

- Flea bugs can be prevented by row coverings. Right after you've planted a row of seeds, you should cover them up.
- Sticky traps, either white or yellow, should be set up every 5-10 m along the rows.
- Larvae's eating is impeded by thick mulch in solitary planting.

Stem or vine borer, *Omphisa anastomosalis*

Damage symptoms

- The larvae eat their way through the main stem and even the storage roots on occasion.
- Infested plants will wilt and die as the larvae chew enormous tunnels and cavities out of the stems.
- When plants are attacked when they are young, they are unable to establish storage roots because the ability to transport food and store it in the roots is severely disrupted.
- To make matters worse, the store roots are rendered unsellable when the larva bores into them.
- It is possible for infestations to reduce crop yields by 30-50% or more during the crop's establishing phase.
- Usually find a clump of frass near the assaulted stem of an infested plant.

Pest identification

- The adult moth measures 15 mm in length and has a 33 mm wingspan; its body is a rusty brown colour, and its white wings are marked with rusty brown spots.
- The larvae can range in hue from pale purple to a yellowish white. It has a brown body and a brown stripe that runs along each side of its head. Adult larvae can reach lengths of 3.0–2.5 cm.

Spread

- The most common ways for this pest to spread are through the use of infected planting material or by the establishment of new fields in close proximity to existing infestations.

Management

- Improved crop hygiene and rotational cropping both help lower pest populations.
- A common method for controlling this pest is to regularly soil up the area around the plant's trunk and roots.
- The sweet potato vine larvae are vulnerable to attacks by earwigs and ants.
- You shouldn't plant sweet potatoes for two growing seasons after a severe insect infestation.
- It is important to keep the region clear of Ipomea weeds so that the pest has nowhere to find nourishment.
- Take care when choosing your plant materials. Do not propagate plants from diseased cuttings.
- Sweet potatoes can be grown successfully under row covers. This will prevent the bug from damaging plant life.

Whitefly, *Bemisia tabaci*

Damage symptoms

- Larvae and adults alike feed on the leaves by puncturing them and sucking out the juices.
- Reduced plant growth is the outcome of the plant's weakened condition and its premature wilting.
- Yellowing, drying, and premature leaf drop can be caused by their feeding,

which could ultimately lead to the death of the plant.

- Honeydew secreted by whiteflies feeds black sooty mould that colonise plant surfaces including leaves and berries.
- Honeydew is enticing to many insects, including ants. Since photosynthesis is inhibited by the mould, the plant fails to thrive.

Pest identification

- Larvae are 0.3-0.7 mm in length, translucent, and oval in shape.
- About 1 mm in length, adults have two sets of white wings and pale yellow bodies. They have a powdered, waxy coating all over their body.

Management

- A fresh crop should not be planted adjacent to an established one.
- Weeds are hosts for the insect, so it's important to maintain good farm sanitation by getting rid of them and other debris near the crops.
- Spread some Neem oil or garlic oil.
- Spray a soapy potato starch mixture.
- Use 10 ml of insecticide for every 4.5 litres of water, such as Admire, Pegasus, Basudin, and/or Vydate L.
- Natural methods of controlling whiteflies rely heavily on parasitic wasps and other predators including predatory mites, ladybird beetles, and lacewings.

DISEASES

Black rot, *Ceratocystis fimbriata*

Symptoms

- The earliest signs of black rot are small, round, slightly depressed, dark brown spots on roots.
- The spots on the tubers grow larger, changing colour from grey to black while wet to a more typical black when dry. Spots include tiny, black fungal structures called perithecia, which have long necks and look like dark bristles. Typically, the rot will remain rigid and quite shallow.
- The flesh beneath the spot turns black if secondary fungus or bacteria infiltrate the tissue, and this blackened patch may extend to the root's centre.
- Discolored tissue may impart a bitter flavour to nearby tissue. There is a risk

that the entire root will rot away. Roots may look fine when harvested, but they may deteriorate while being stored, transported, or sold.

Survival and spread

- Between sweet potato plantings, the earth is home to a persistent fungus.
- The inoculum can be obtained from contaminated soil or tubers.
- Infectious diseases can spread from the seedbed to the field and spread among the plants.
- Sweet potatoes in storage are especially susceptible to the rapid spread of the disease.

Management

- All sweet potato crop leftovers must be eliminated right after harvest.
- Intercropping with crops including cabbage, ginger, yam, corn, lettuce, brinjal, and cucumber (2-3) years apart is recommended.
- When a sterile location is not accessible, disinfect the seedbeds.
- The best way to propagate plants is by cuttings of healthy stems.
- Harvested roots should be cured promptly (Cure roots between 85–95 °F and 85–90% relative humidity for 5–10 days).
- Roots with black rot should not be washed or packaged.
- Any tools that have come into contact with an infected crop should be disinfected.
- Increasing the temperature of Thiabendazole and Benomyl dips used to treat seed roots and slips before to planting resulted in significant improvements in both germination and disease resistance.
- Ferbam dips at 44-54 °C were similarly successful in preventing black rot.

Rhizopus soft rot, *Rhizopus stolonifer*

Symptoms

- Root end infections and deterioration are very prevalent, though infections can start anywhere on the root.
- To some extent, rotting can be slowed by a lack of moisture, but when the air is damp, sweet potatoes become mushy and watery and their entire roots rot away in a matter of days.

- Sweet potatoes become significantly "whiskered" with a greyish black fungal growth if the humidity is too high. What sets Rhizopus soft rot apart from other types of storage rots is this specific trait.
- There is no noticeable change in root colour, but a smell is created that draws fruit flies.
- When the relative humidity is between 75% and 85%, the risk of contamination increases significantly. The longer roots are kept underground, the more vulnerable they become.
- Sweet potatoes are especially vulnerable to contamination due to chilling and heat damage. When sweet potatoes are washed, packed, or delivered to market when it's chilly, soft rot can cause serious damage.

Favorable conditions

- The infectious process can only begin with a physical injury, specifically a bruise.
- Infections spread more quickly at warm temperatures (25°C to 29°C).
- Increased sporulation of *R. sporogenes* occurs at high relative humidities (above 90%). increased secondary inoculum due to stolonifer.

Spread

- Diseased soil, airborne spores, stored infected roots, and contaminated containers, baskets, or implements all contribute to the spread of this disease.

Management

- Always use caution when harvesting sweet potatoes to avoid injuring the plants.
- When it comes to preventing soft rot, this is by far the most effective strategy.
- Immediately following harvest, cure roots in the correct manner.
- Roots should be kept at 13 °C and 85% relative humidity in a well- ventilated cold room.
- Never touch a root you've stored away because doing so could induce an infection.
- After harvest, use a fungicide.
- When growing sweet potatoes, it's important to keep them out of direct sunshine for long periods of time (to avoid heat damage) and also to keep them from getting too cold in the field.

Charcoal rot, *Macrophomina phaseolina*

Symptoms

» Waterlogged brown to black lesions form on stems and petioles in the field. Vine tips may wilt if the stem becomes waterlogged and gives way.

» Sometimes an entire plant will die, other times only one or two tendrils may fall.

» Sweet potatoes can be lost in storage due to the fungus known as charcoal rot. However, this usually only happens in small amounts. Similar diseases, such as black rot and Java black rot, are often misdiagnosed as this one.

» Sweet potatoes can rot in storage, first manifesting as a reddish brown to brown, firm, moist rot right under the skin.

» The decaying sweet potato's virus spreads inward as the rot worsens. Within the inflamed tissue, two different areas develop.

» There is a continuous reddish-brown decay at the leading edge, and a black zone develops behind the zone of active decay.

» Charcoal rot can sometimes be contained within small areas, but in most cases it spreads throughout the root and leaves it hard and mummified.

Favorable conditions

» Disease appears to flourish at high temperatures (84-88 °F).

Survival and spread

» Fungi can live in dead plant matter or in the soil for long periods of time.

Management

» When harvesting, take care that the tubers don't get harmed. Any that are broken must be thrown away.

» Curing sweet potatoes properly right after harvest can greatly reduce the amount of charcoal rot that develops on them.

» Tubers should be kept in a dry, cool, and dark place once they have finished curing.

Cercospora leaf spot, *Cercospora ipomoeae, Pseudocercospora timorensis*

Losses

» It has been anticipated that yields will drop by 15-20%.

Symptoms

» Spots first appear yellowish brown and then darken over time as the disease progresses; their borders may be round, ovoid, or irregular.

» The spots on the leaf eventually join together to form larger patches that spread across the entire leaf.

» Leaves with more age have more spots.

» Under ideal conditions, significant examples of shot hole and defoliation are frequently visible.

» Water causes the most serious damage to leaves.

Favorable conditions

» Tropical regions with high temperatures and high humidity levels are the most common habitats for this species.

Spread

» Certain weed species serve as hosts for the pathogen during the winter.

» The spots on the leaves produce a lot of conidia, which are then carried by the wind and splashing rain.

Management

» Only plant good soil and plant stuff.

» Treatment with 0.25 percent Mancozeb (Dithane M-45) sprayed three times at 2-week intervals decreased disease intensity (19.89 percent compared to 50.7 eight percent in the control) and increased tuber output (21.76 tonnes per hectare compared to 12.91 tonnes per hectare).

Alternaria leaf spot/Leaf and stem blight, ***Alternaria bataticola***

Symptoms

» Target-shaped lesions of varying shades of brown appear on the leaves, petioles, and stems.

» Older leaves are typically affected, and lesions may have a yellow halo around them. Oval, small, greyish-black lesions with lighter centres can appear on the stems, leaf petioles, and, rarely, the leaves themselves.

» Girdling of the stem occurs when lesions on the stem and petiole grow and eventually join together.

» These spots can cluster, causing the plant to lose its leaves and eventually die from suffocation at its base.

» Infected leaves eventually turn yellow and drop off.

Favorable conditions

» Conditions ideal for the spread of disease include high humidity.

Survival and spread

» Mycelium or conidia allow fungi to live on in plant matter and soil.

» Airborne spores, splash water, irrigation water, wind, insects, and rain all play a role in the disease's rapid proliferation.

» The disease can be transmitted across great distances on cuttings that are then planted.

Management

» All sweet potato crop debris must be removed and disposed of shortly after harvest.

» When possible, use sweet potato cultivars that are resistant to disease (NASPOT 3, 5, 6, and 11) or tolerant (Kakamega and Ejumula).

» It's important to use only disease-free seeds.

» It's best to not water the lawn with a sprinkler gun.

» Fungicides such as Mancozeb, Chlorothalonil, and Dyrene should be sprayed.

Fusarium wilt, *Fusarium oxysporum* f. sp. *batatas*

Symptoms

» Vessel weakness is the result of the fungus's attack.

» Young, healthy leaves will turn yellow first, signalling the beginning of the disease.

» Inner stems darken as the vascular system turns black.

» Between individual stems, this shadowing may reach several feet.

- Dying plants fall and wilt after infection.
- A blackened ring appears around 6 mm beneath the epidermis of infected tubers, and these grow a surface rot with shallow, sunken, circular lesions while in storage.
- The risk of disease transmission to plants grown from these tubers is high.

Favorable conditions

- When soil moisture is below 50% and temperatures are about 30 °C, infection is most likely to occur.

Survival and spread

- Infected tubers and plants, irrigation water, and tools are major vectors of spread.
- The fungus can survive in the soil for years on dead plants and spread locally via rain, wind, animals, vehicles, and other means.

Management

- Plant material should be disease-free.
- Keep away from areas that have had wilt in the past.
- After harvesting, remove infected plant materials and dispose of them properly.
- Integrate sensible crop rotation practises into your farming practises.
- Dip treatments using the systemic fungicide Thiabendazole have showed some promise.
- It has been reported that the cvs 'Pusa Safed' and 'Kalmegh' are resistant to wilt.
- Inoculating the cuttings with a non-pathogenic strain of *Fusarium oxysporum* just before they are planted in the field.

Soil rot, *Streptomyces ipomoeae*

Symptoms

- Dwarfed plants caused by a severe infection will only grow one or two little, pale green vines.
- These signs are caused by trauma to the plant's root system.

» A lot of the lateral roots get hacked off, and the ones that survive often become twisted or otherwise mutilated in the process.

» The most noticeable indicator is the appearance of pits on the tubers, which initially appear as damp, dark areas.

» The tubers' characteristic pits with jagged sides are formed as the overlying skin splits and necrotic tissue underneath is shed as the spots expand.

» There is a wide variety in size between the pits, from 5 mm to 30 mm.

» Lesions may encircle the tubers, preventing growth in that location, but the tubers will continue to expand on either side of the lesion.

Survival and spread

» The organism responsible for soil rot has a long life cycle in the soil, and alkaline or barely acidic soils exacerbate the condition.

» The pathogen is disseminated via infected tubers, which are transported from infected areas, by infected individuals, and by infected animals.

Management

» Preventing the spread of contaminated materials (soil and tubers) to unaffected fields is the most efficient technique of controlling this disease.

» Plants and seeds should be in good health.

» For as long as feasible after a disease has been confirmed, other crops can be grown in the infected field.

» Banrot, Rizolex, Captan, or Mankocide are suggested for efficient management.

» Sulfur, when added to the soil, acts as an acidifier and reduces disease.

Bacterial soft rot, *Dickeya* (*Erwinia*) *chrysanthemi*

Symptoms

» Soft rot, which causes roots to become light brown and liquid, can happen in the field or, more frequently, in storage.

» The edge of lesions on storage roots typically appears dark brown. Though their exteriors may look healthy, some storage roots actually have rotted away on the inside.

» Roots that are infected will develop dark streaks in the vascular tissue and eventually rot away softly and wetly.

- Oftentimes, mother roots will rot in plant beds. Waterlogged brown to black lesions form on stems and petioles in the field. Vine tips may wilt if the stem.
- Becomes waterlogged and gives way.
- Sometimes an entire plant will die, other times only one or two tendrils may fall.

Favorable conditions

- Wet and warm conditions are ideal for the spread of the disease.
- At a temperature of 30 °C, the spread of disease is at its most rapid.

Survival and spread

- It can be found in the soil around the roots of crops, in the crop detritus, and in weeds.
- Planting material, irrigation water, equipment, animal grazing, worker shoes, etc. can all be vectors for the spread of the disease.

Management

- Making use of vine cuttings collected from the very top of the plant.
- Caring for the storage roots properly after harvest can prevent injuries.
- Change to non-host crops every year for a minimum of three years.» Choose disease-free pastures to dig up your mother roots.
- Roots with infections should be discarded.
- For successful transplanting, use vines that have been clipped above ground.
- Avoid submerging sweet potatoes in water as much as possible while handling them.

Feathery mottle disease, *Feathery Mottle Virus*

Symptoms

- Some varieties have the characteristic irregular chlorotic patterns (feathering) along leaf veins, as well as faint to obvious chlorotic patches with or without purple edges.
- The leaves frequently develop irregular mottled patches, which, in some cultivars, are surrounded by a purplish halo.
- The veins of some leaves occasionally take on a dingy, feathery appearance.

- » Storage roots become skinny and elongated, and plant vitality decreases.
- » A severe infection can cause a lot of tiny fissures all over the root surface and light flesh, especially in orange kinds.
- » The rod-shaped sweet potato feathery mottle virus can bend and twist.

Transmission

- » Numerous aphid species, including *Aphis gossypii*, *Aphis craccivora*, and *Myzus persicae*, are capable of transmitting SPFMV in a non-persistent manner through the sap they feed on.
- » The virus is kept alive through vegetative reproduction.

Management

- » Take samples from thriving gardens.
- » Cultivate plants that can withstand the elements. To a certain extent, "Namujuna" is immune to the SPMMV.
- » Rotating your crops is a smart farming practise.
- » When taking cuttings, make sure to use clean or sterilised instruments.
- » Keeping new plants separate from established ones is essential.
- » Eliminate as many vector insects as possible.
- » Through a combination of meristem-tip culture and thermotherapy, virus-free plant tissue has been obtained.

References

Ray, R. C., Ravi, V., Hegde, V., Rao, K. R., & Tomlins, K. I. (2010). Post harvest handling, storage methods, pest and diseases of sweet potato. Sweet Potato: Post Harvest Aspects in Food, Nova Science Publishers, Inc., New York, 27-57.

Lenné, J. M. (1991). Diseases and pests of sweet potato: south-east Asia, the Pacific and East Africa (NRI Bulletin No. 46).

Jansson, R. K., & Raman, K. V. (2019). Sweet potato pest management: a global overview. Sweet Potato Pest Management, 1-12.

Traynor, M. M. (2006). Sweet Potato: Production Guide for the Top End. Department of Primary Industry, Fisheries and Mines.

20 Taro / Colocasia

INSECT PESTS

Armyworm, *Spodoptera litura*

Damage symptoms

- Initially the larvae stay clustered together, but they eventually spread out from the egg mass, removing the interveinal leaf surface and skeletonizing the leaves as they go.
- Petioles are consumed alongside the rest of the leaf by later (solitary) stages.
- The older larvae only feed at night.
- When armyworm populations are high enough, they can defoliate a crop by eating big sections of leaves.
- When this happens, masses of larvae move from one area to another, looking for a new source of nourishment.

Pest identification

- Mature larvae range in colour from dark green to brown and are characterised by brilliant yellow stripes along their back.
- The moth has a brownish green colour and measures 15-20 mm in length and 30-40 mm in wingspan.

Management

- To prevent further spread, infected leaves are discarded.
- The adult moth can be deterred from the taro field by planting *Coleus blumei* (pate) around or within the crop.

- To remove severe pest problems, Malathion spray is usually sufficient.
- Parasitoids like the flies *Palexorista* sp. and *Telenomus* sp. (Hymenoptera) feed on caterpillar eggs, whereas the flies *Apanteles* sp., *Euplectrus* sp., and *Zele* sp. prey on caterpillar larvae.
- *Bacillus thuringiensis* subsp. *kurstaki* (Dipel) has been used to combat *S. litura.*
- The soil Actinomycete *Saccharopolyspora spinosa*, which is used to create the pesticide spinosad, is also effective.

Taro beetles, *Papuana woodlarkiana, P. biroi, P. huebneri, P. trinodosa*

Damage symptoms

- Tunnels are created by adult beetles as they devour taro corms buried underground.
- The damage caused by feeding is significant since export markets will not take crops with any damage, and local markets would not accept crops with more than 15% damage.
- There is a risk that the corms have been damaged beyond repair, making them unfit for human or animal consumption.
- The beetle may cause the death of young plants by eating away at the developing shoot, while older plants may grow more slowly and lose some or all of their leaves.

Pest identification

- The average length of a beetle is 25 mm, while its width is 12.5 mm.
- Males have a horn that protrudes from the top of their heads. Miniature forms of these characteristics can sometimes be found in females.
- During the first few months, the body is a glossy dark brown.

Favorable conditions

- This beetle is most at home in rich, organic loams or silt loams that retain water.

Management

- The use of wood ash as a fertiliser.
- Crop rotation, using sterile planting media, and the elimination of potential breeding grounds in agricultural areas are all examples of cultural methods of pest control.

- Applying Imidacloprid/Bifenthrin Spray.
- *Metarrhizium anisopliae*, a harmful fungus, is being sprayed in a field.

Aphids, *Aphis gossypii*

Damage symptoms

- If there are a lot of aphids and there isn't much moisture, the leaves will wilt much sooner than usual.
- The plants wilt and may even become stunted if the condition persists.
- Honeydew, which the aphids secrete, can cause further damage.
- Honeydew provides a food source for sooty moulds, which then colonise and blacken the leaves, inhibiting photosynthesis and weakening the plant.
- The *Dasheen Mosaic Potyvirus* is spread by aphids.

Pest identification

- Nymphs resemble adults in appearance but are smaller and have no wings. Nymphs in their fourth stage are about 1 millimetre in length.
- The length of adults is between 0.9 mm and 2.5 mm. Adult aphids, which lack wings, can be anywhere from a bluish-green to a deep emerald hue. The winged varieties are often darker overall, with a black head and torso and a yellowish-green abdomen with dark-brown veins.

Management

- Colocasia should not be planted in close proximity to alternate host plants including melons, cotton, cucumbers, and other cucurbits.
- Insecticides such as Carbaryl WP should be sprayed at a rate of 1 kg per 450 L (high volume) or 112-168 L (low volume) of water, while Diazinon EC 11 ml per 13.5 L of water, Dimethoate EC 4.4 ml per 13.5 L of water, and Malathion EC 14.2 ml per 13.6 L of water should be applied in similar fashions.
- *Ischiodon scutellaris* is a voracious eater that preys on aphids and other aphidophagous predators like *Verania discolour*, *Verania inops*, and *Pseudospidimerus circumflexus*.
- Parasitoids, four types *Aphidencyrtus aphidivorus* (Encyrtidae) and *Aphelinus mali*, *Aphelinus* sp., and *Coccophagus cowperi* (Aphelinidae) feed on *Aphis gossypii*.

Whitefly, *Bemisia tabaci*

Damage symptoms

- All life stages, from adults to nymphs and pupae, can be found on the underside of a leaf.
- Adults and juveniles alike feed on the sap of infested plants, hastening their demise by making the leaves turn yellow and fall off.
- While whitefly can promote wilting during drought, symptoms this severe are unusual in taro due to the low numbers of whitefly needed to cause direct harm.
- The accumulation of sooty moulds on honeydew deposits is an example of indirect damage, but it is not something that typically occurs.
- Colocasia may experience retarded development.

Pest identification

- Insect nymphs are the size of pinheads and have a golden, elliptical, scaled appearance.
- The adult is about 1 mm in length and has a wingspan of about 2 mm. It has a yellow body and white waxy wings with broad yellow tips.

Management

- Establish hedges between fields to prevent whiteflies from spreading from one crop to another and to attract predatory insects.
- A heavy downpour or sprinkler watering can help control the bug population.
- Apply pesticides such as Carbaryl WP 34 g, Malathion WP 14 g, or Trichlorfon WP 23 g a.i./13.6 L of water.
- Use soap, oil, and systemic insecticides to remove the pupae (*puparium*).
- You should encourage predatory mites (*Amblyseius* spp.), parasitic wasps (*Encarsia formosa, Eretmocerus*), ladybird beetles, and lacewings.
- Elimination of harvest waste through destruction stops accumulation and further dissemination.

Plant hopper, *Tarophagus proserpina*

Damage symptoms

- It's common to find plant hoppers, which are sap-sucking insects, on the petioles and the undersides of leaves. When leaf hoppers feed, they penetrate the plant, causing a reddish crust of sap exudate to form on the outside of the leaf stalks close to the ground.
- When a plant has a lot of pests, it may not grow as well or even die.
- *T. proserpina* is a persistent vector for two devastating taro viral diseasees; once infected, the plant hopper can spread the disease indefinitely.

Pest identification

- Most nymphs are very pale in colour.
- The adult forms possess a brown and white coloration.

Spread

- Planting tiapula with plant hopper eggs implanted in the petioles and the long- winged variety are both effective means of long-distance dispersal.

Management

- Delaying the commencement of an assault may be possible by first planting tiapula in fields that are not affected with plant hoppers.
- The corm, together with the attached petiole, must be cut off at both ends.
- Switching to non-host plants for your crop rotation.
- Taro can be protected from plant hoppers and armyworms by planting it with *Coleus blumei* or pate.
- The mirid *Cyrtorhinus fulvus* is an effective egg predator for managing plant hopper populations.
- Imidacloprid (Confidor, Mustang), Malathion, Carbaryl, Diazinon, and Acephate are effective pesticides but they shouldn't be used unless all other options have failed.

Hornworm, *Hippotion celerio*

Damage symptoms

- Leaves on infected plants are often missing big sections and have a ragged appearance.

- Caterpillars eat quite quickly.
- When hornworm populations are high, the plants' leaves are eaten off.
- An infestation of caterpillars can cause significant defoliation.
- During the day, the leaves and the larvae inside them are visible.

Pest identification

- Adult larvae have a body length of 8-9 cm and range in colour from a medium brown to a dark brown.
- Wingspan for a hawk moth ranges from 4 to 9 centimetres. The size of their head and eyes make them stand out.

Management

- The larvae are noticeable and large enough to be plucked by hand from the plants.
- This is the most effective method of management for small-scale taro gardens.
- Spraying pesticides on infested areas could reduce hornworm populations.
- Imidacloprid (e.g., Repel), Spinosad (e.g., Success), *Bacillus thurungiensis* (e.g., Delfin, Thuricide, Dipel), and Indoxacarb (e.g., Steward) are just some of the insecticides you can use (e.g., Confidor, Mustang).
- In the wild, a Hymenopteran called *Charops hersei* commonly parasitizes hornworm caterpillars (between 6-10%). In the wild, the wasp *Polistes* sp. also feeds on *Hippotion* spp. caterpillars.

White spotted flea beetle, *Monolepta signata*

Damage symptoms

- Worms feed on small plant roots and root hairs.
- Bugs like beetle's chew holes in leaves.
- Lose yield by damaging the lamellae of the leaves.
- Extreme infection might cause total demise of young plants.

Pest identification

- Adult beetles are around 3–4 millimetres long and dark reddish brown in colour.

- Two large white specks on brown elytra.

Management

- Use a 0.1% Carbaryl spray.

Spiraling whitefly, *Aleurodicus dispersus*

Damage symptoms

- When whitefly populations are very high, the insects pierce leaves and feed on the sap, killing plants prematurely.
- Honeydew and the waxy, white, fluffy, woolly substance produced by whiteflies cause indirect injury.
- Black sooty mould, which feed on honeydew, reduce plant health by inhibiting photosynthesis.
- As the name suggests, it lays its eggs in characteristic spiral patterns on the undersides of leaves.

Pest identification

- The underside of the leaves is covered in a white spiral pattern of eggs and waxy material, making both the immature stages and the adults stand out.
- The eggs are 0.3 mm in size and have a smooth, oval form that ranges in colour from pale yellow to tan.
- Waxy tufts appear on the larvae as they develop.
- Males reach a mature size of 2.2 mm and females of 1.7 mm.

Management

- Quarantine methods that work to stop the spread of disease around the world.
- Containing biological threats during quarantine breaches.
- Spiral whitefly populations have been drastically reduced everywhere the parasitic wasp *Encarsia haitiensis* has been introduced.

DISEASES

Leaf blight, *Phytophthora colocasiae*

Symptoms

- When first noticed, lesions appear small, dark, and round. However, they quickly grow to be between 2.5 and 5 cm in diameter, changing from a dark purple to a brownish colour with a chlorotic halo around the spots.
- Spots eventually join together to form yellow or brown rings with white fluff on both sides of leaves as the disease spreads.
- As the lesions on a leaf dry and fall off, holes may develop.
- Corm rot after harvest manifests as huge patches of dark grey to blue decay with hazy borders.
- At the first sign of disease, spray with 0.25 percent Mancozeb; 20 days later, spray with 2 percent Ridomil; this is the standard protocol for treating tuberculosis.
- Telia seed tubers were treated with *Trichoderma viride* (5 g/kg cow dung slurry), and 1 kg of compost supplemented with *T. viride* was applied to each planting hole.

Survival and spread

- Infected tubers and weedy Colocasia plants can harbour the pathogen for years.
- Spots on the leaf surface release spores, which are easily dispersed by splashing rainwater.

Favorable conditions

- Conditions that include regular rain, cloudy skies, and low nighttime temperatures are ideal: 20-22 °C during the night and 100% relative humidity (RH) during the day.

Management

- Plants are used that are in good health.
- Cultivate resistant strains like Muktakeshi and Jankhri.
- Planting disease-resistant crops early (in May) with early-maturing varieties such as Telia.

- Clean up the field and remove the diseased leaves.
- Crop rotation that includes non-host species, such as maize.
- Mancozeb (a topical) and Metalaxyl (a systemic) preventative sprays.
- *Trichoderma viride*, *Trichoderma harzianum*, *Bacillus subtilis*, and *Pseudomonas fluorescens* are used as biocontrol agents.
- *Coleus blumei* is a good plant to plant in taro fields to encourage the growth of predator insects.

Corm soft rot, *Pythium* spp.

Symptoms

- The symptoms include a decrease or complete disappearance of the root system, as well as stunting and leaf yellowing.
- There is a risk that immature plants will be killed or stunted if they are assaulted before they have produced any leaves.
- Corm rot typically manifests at the corm's base and spreads upwards, threatening to destroy the entire corm.
- Firm and varying in colour from yellowish white to grey, blue, or purple, rotting tissue is easily recognisable.
- Along with the corms, the roots rot away.

Favorable conditions

- Lots of water, warmth, and poor soil are all positive factors.
- Wetlands and regions with poor drainage are especially vulnerable.

Survival and spread

- The pathogen can live in the soil for years, making its way from plant to plant.

Management

- Wide rows, tall mounds, timed planting, disease-free seedlings, ridge planting, crop rotation, and organic fertiliser use are all examples of cultural management.
- Inoculating a disease-free field with disease-free material.
- Use of a lot of mulch and compost to enrich the soil.
- Early detection and treatment of plant diseases requires prompt removal

of infected plants and soil.

» Quick picking at the peak of maturity.
» It's important to clean up the fields after harvesting.
» Copper ox chloride, often known as Captan, is dipped into the soil and then poured over the plants.
» Carbendazim (0.1%) with Streptomycin treatment prior to harvest (1,000 ppm).
» Use of Copper Oxychloride (0.3% solution) and Streptomycin after Harvest (100 ppm).
» Seed tubers are treated with *Trichoderma viride* (0.5%) before being stored in pits enclosed with polythene sheets containing damp sand.

Brown leaf spot, *Cladosporium colocasiae*

Symptoms

» Brown leaf spot (also known as ghost spot) is a disease that affects older leaves and appears as reddish-brown dots or blotches with diffuse edges and often dark, diffuse centres.
» Subsurface hyphae, sporophores, and conidia of the fungus cause the dots on the underside of the leaf to turn a dark brown colour.
» Typically, the spotting is not as noticeable on the underside of the leaf.
» Spots can have a yellow halo around them or a hazy brown border.
» When there are multiple spots on a single leaf, the spots tend to be smaller than the maximum size of 15 mm.

Management

» To lessen the spread of the disease, contaminated leaves should be picked up and burned.
» Sprays of fungicides used to combat leaf blight will have the same effect on brown leaf spot.

Corm rot, *Athelia rolfsii*

Symptoms

» At the soil's surface, at the base of the petioles, a soil-borne fungus infects taro, causing the corms and roots to rot and the leaves to wilt.

- White mycelia spread out like fans over the affected area, sometimes even colonising neighbouring organic substances.
- Wilting leaves are a common problem.
- Symptoms of the condition include an abnormally high number of fallen leaves.
- Sclerotia ranging in colour from white to a reddish brown, measuring between 1 and 2 millimetres in diameter, are frequently seen near the site of infection.
- After harvest, the corms become infected with the fungus through the wounds left by the severed suckers, resulting in a pinkish rot.
- Colocasia that has matured too much may be more susceptible to the disease.

Survival and spread

- The lifespan of a sclerotia can extend into the decades.
- Since the pathogen is so widespread, inoculum sources are not hard to get by

Management

- Taking out and burning diseased plants.
- Using cultural techniques that discourage fungal growth, such as deep ploughing and fostering the development of microorganisms (particularly *Trichoderma* spp.), is essential.
- Solarizing the soil by covering it with plastic and letting the sun warm the air underneath.
- Assisting the soil by adding lime
- Drenching the soil with Quintozone or Dicloran has been suggested as a method of management.
- Planting non-host or tolerant crops in between outbreaks can drastically cut the spread of disease. The cereal family (Poaceae) of plants is resistant to the fungus.

Mosaic disease, *Dasheen Mosaic Virus*

Symptoms

- Some plants have dazzling white or yellow featherlike patterns along the veins and sometimes across the leaf blade, while others have small, irregular, and scattered patches of grey, green, yellow, and sometimes white along or between the principal veins.

- Leaves that are abnormally tiny, deformed, and small on plants.
- For certain leaves, the blades have been removed, leaving only the strap-like structures.
- It's not uncommon for these plants to perish.
- The infection causes a feathery mosaic pattern around the veins.
- It's also possible to see the petioles are streaked with a yellowish-green colour.
- Petioles are short and the leaves of seriously affected plants are little and don't unfold normally.
- A single viral particle is a long, flexible rod that is around 75 nm in length.

Transmission

- *Myzus persicae, Aphis gossypii*, and *Pentalonia nigronervosa* are aphid species that spread the virus in a short-lived, intermittent cycle.
- As with many viruses, this one can be spread through contact with sap.

Management

- There needs to be a strict quarantine in place.
- All transplants should be of pathogen-free, tissue-cultured plantlets, according to the guidelines.
- Planting only approved healthy planting material.
- Meristem tip culture can be used to get healthy cuttings for planting.

References

Plucknett, D. L., De la Pena, R. S., & Obrero, F. (1970). Taro (Colocasia escalenta). In Field Crop Abstracts (Vol. 23, No. 4, pp. 413-426).

Jackson, G. V. H., & Gollifer, D. E. (1975). Disease and pest problems of taro (Colocasia esculenta L. Schott) in the British Solomon Islands. PANS Pest Articles & News Summaries, 21(1), 45-53.

Gollifer, D. E., Jackson, G. V. H., Dabek, A. J., & Plumb, R. T. (1978). Incidence, and effects on yield, of virus diseases of taro (Colocasia esculenta) in the Solomon Islands. Annals of Applied Biology, 88(1), 131-135.

Revill, P. A., Jackson, G. V. H., Hafnerc, G. J., Yang, I., Maino, M. K., Dowling, M.

L., ... & Harding, R. M. (2005). Incidence and distribution of viruses of taro (Colocasia esculenta) in Pacific Island countries. Australasian Plant Pathology, 34, 327-331.

Morishita, M. (1988). Taro (Colocasia esculenta Schott.). In Crops II (pp. 322-338). Berlin, Heidelberg: Springer Berlin Heidelberg.

Okonkwo, C. A. C. (1993). Taro: Colocasia spp. In Genetic Improvement of Vegetable Crops (pp. 709-715). Pergamon.

21 Tea

INSECT PESTS

Shot-hole borer, *Euwallacea* (*Xyleborus*) *fornicatus*

Damage symptoms

» The tiny pinholes that lead into the tunnels, which are lined with a fungus, are a telltale sign of an infestation (*Fusarium bugnicourtii*).

» Male beetles feed on the fungus *Fusarium bugnicourtii* that grows in the galleries made by the female beetles, which causes branch breakage and, ultimately, crop loss.

» Discoloration occurs around the bullet holes in infected branches, and leaves on the end of the branch turn yellow, droop, and fall off.

» Between the months of October and December, the pest population explodes.

Pest identification

» Black female beetles (length of body 2-2.5 mm) are highly sclerotized.

Management

» Trimming to a medium degree in infected areas.

» Don't do any sort of pruning if the weather is too dry.

» There should be two applications of the specified pesticide, one in April and one in December, sprayed after pruning is complete.

» This is the year that you should apply a 1:2 mixtures of nitrous oxide and potassium oxide.

- Eliminate the affected tree limbs.
- *Beauvaria bassiana* (sold as the wettable powder 'Bio- Power' formulation) should be sprayed twice, at a rate of 1.5 kg/ha, between the months of May and October, when humidity is at its highest.
- Infested plantations in their third and fourth years can benefit from 400 stems of chopped *Montonoa bipinnatifida*, 22 to 28 days old, approximately 90 cm in length and 25 to 30 mm in thickness.

Mosquito bug, *Helopeltis theivora*

Damage symptoms

- Adults and juveniles alike will use their rostrums, which look like tiny needles, to pierce the plant tissues and syphon the sap from buds, new leaves, and young, delicate stems.
- The intense eating that causes punctures to form causes the skin to turn a rusty brown colour.
- Curling, progressive deformation, and reduced leaf size are all symptoms.
- In the face of a major incident, shoots dry up and crop loss is nearly complete.

Pest identification

- The form of an egg resembles a lengthy sausage.
- The nymph's dirty yellow legs are long and delicate.
- The adult has greenish brown wings and a greenish brown body with black and white markings.

Management

- Reducing the prevalence of this bug requires intensive stalk removal during plucking.
- Helopeltis thrives in weeds because they provide ideal cover and food sources. Planting weeds and wild host plants including *Mikania cordata, Bidens bifernata, Emillia* sp., *Polygonum chinense*, and *Lantana camera* in and around tea farms has been shown to lower insect populations.
- Tea mosquito eggs are destroyed by cutting off the stalks where they developed.
- A number of other insecticides are suggested for use, including Endosulfan 35 EC at 1000 ml/ha, Quinalphos 25 EC at 750 ml/ha, Chlorpyrifos 20

EC at 750 ml/ ha, Fenthion 80 EC at 200 ml/ha, and Quinalphos 25 EC + Dichlorvos 76 EC at 750 + 250 ml/ha.

» *Mymarid, Erythmelus helopeltidis*, parasitizes between 52 and 83 percent of mosquito insect eggs.

» When introduced at 500-1000 adults/ha, *Chrysoperla carnea* consumes mosquitoes as a primary food source. Both *Chrysoperla zastrowii sillemi* and *Mallada* sp. The nematodes (*Hexamermis* sp.) and the fungus *Beauveria bassiana* are the main enemies.

» The use of Biogreen-5 resulted in a statistically significant (89.06%) decrease in tea mosquito populations (*Trichoderma viride, Pseudomonas fluorescens, Beauveria bassiana, Metarrhizium anisopliae* and *Bacillus thuringiensis* based consortial bioformulation).

Green scale, *Coccus viridis*

Damage symptoms

» Some types of scales are able to cling to stems in addition to the leaves.

» Puncture plant tissue and drink the sap to survive.

» The plant is weakened by their feeding.

» Yellowing of the upper leaf surface, reduced flower size, leaf drop, twig dieback, and ultimately plant death might result from a severe infestation.

Pest identification

» Females can be a range of colours and shapes, from dark brown or grey to virtually black. Typically, male are white. They're tiny, unable to move, and sporting no discernible legs.

» Young insects and spiders are a vivid yellow.

Management

» Scales can be easily scraped off a plant with a mild infestation and thrown away.

» Hand plucking and destroying infected leaves is highly effective if just a small number of leaves are affected.

» Sticky cards, tape wrapped around a branch, or a bag containing a shoot or leaf infested with crawlers can be used to track their emergence.

» When the plants have done blooming and the threat of cold weather has gone in the spring is the best time to spray with refined Horticultural oil (Bonide All Season Spray Oil, Ferti-lome Scalecide, Green Light Horticultural Oil Spray, or Ortho Volck Oil Spray). Spray twice in a 10-day span between 40 and 85 °F.

» Acephate, Malathion, Cyfluthrin, and Carbaryl are only some of the insecticides approved for use on tea scale crawlers.

Phassus borer, *Sahydrassus malabaricus*

Damage symptoms

» Young caterpillars consume the bark and sapwood, causing cankers to form on the stems and stout branches.

» Caterpillars excavated tunnels that went all the way down to the roots.

» Chewable wood and silk are used to seal off the entrance.

» It can go both forward and backward without difficulty.

» A frassy mat, made of powdered wood and silk, is hung next to the holes of attacked plants, indicating their location.

Pest identification

» Large and white with a brownish tint, the adult moth is quite distinctive.

» The adult has pale grey mottling on its forewings.

» Fat caterpillars are the larvae.

» Larvae of this species reach maturity at a length of 6-10 cm and have a yellowish hue.

Management

» Cut down and dispose of any dead or dying trees and larvae.

» To attract and kill adults, set up a light trap at a rate of 1/ha.

» Find the caterpillar's living hole, and then spike it with an iron hook to kill it.

» Fill the hole with a ten millilitre solution of monocrotophos and ten ml of water, and then plug it with mud. After fruit has been harvested, inject the stems with a growth hormone. Wait 40 days before harvesting again.

» Effective pest control can be achieved with the use of Endosulfan or Quinalphos spray.

Thrips, *Scirtothrips bispinosus*

Damage symptoms

- Damages plants by feeding on young, vulnerable sections above ground, leaving behind feeding scars, distorted leaves, and discoloured buds.
- Infected leaves roll upwards, become crumbly, and fall off the plant.
- Buds that have been infected by this pest get dry and fall off.
- The sucking markings run in a straight line parallel to the major vein on the underside of the leaf.
- Leaves curl and exhibit feeding markings as parallel lines on both sides of the midrib.
- A preference for new growth and buds.
- Streaks of bleeding are the result of lacerations caused by constant eating.
- A curled and irregular leaf surface is the result of a physiological change in the leaf.
- Fields that are still recovering from a pruning attack are especially vulnerable.
- Causing crop damage because of excessive waiting time before tipping.

Pest identification

- Freshly deposited eggs have essentially no colour and have the shape of a bean, narrower at one end.
- The nymph is an orange-yellow colour during its second stage of development.
- As an adult, its body is a pale yellow, with a lighter belly.

Management

- Thrips are more likely to attack in regions with little or no cover. Thrip populations can be kept in check with just careful attention to shading.
- Yellow sticky traps and consistent shading are used to count the population in the field.
- Anthocorids (*Anthocoris* and *Orius* spp.), predatory thrips (*Aeolothrips intermedius* and *Mymarothrips garuda*), *Chrysoperla carnea*, and *Mallada* spp.
- To remove thrips, release 500-1000 *Chrysoperla carnea* adults per hectare.
- *Verticillium lecanii*, an entomopathogen, should be applied as a spore suspension at a rate of 1.5 kg/ha in the evening, when humidity is higher. The

fungi *Paecilomyces fumosoroseus, Beauveria bassiana, Metarrhizium anisopliae*, and *Steinernema* sp. are all useful pathogens.

» The following chemicals were suggested for the management of thrips: Quinalphos 25 EC at 750 ml/ha, Dimethoate 20 EC at 500 ml/ha, Endosulfan 35 EC at 1000 ml/ha, and Quinalphos 25 EC + Dichlorvos 76 EC at 750+250 ml/ha.

Red spider mite, *Oligonychus coffeae*

Damage symptoms

» The upper leaf surface is infested first, starting at the midrib and spreading out down the veins.

» The upkeep foliage becomes a rusty bronze colour as a result of feeding, and infested fields can be easily identified from far away.

» The leaves fall off the trees because of the severe infestation.

» Silken strands are woven by mites into a web on the leaf.

Pest identification

» Female adults are oval in shape and vary in colour from a bright crimson front to a dark purplish brown back.

» The eggs are a dark crimson colour and are round and equipped with a little filament.

Management

» Controlling shading and weed growth, as well as eradicating potential host plant competitors (*Bidens, Ageratum, Conyza, Crassocephalum*, etc.), are essential.

» Fenpropathrin 10 E 500 ml/ha, Dicofol 18.5 EC 1000 ml/ha, Ethion 50 EC 750 ml/ha, Lime Sulfur 1:40, Dicofol 18.5 EC 1000 ml/ha, Dicofol + Ethion 500 + 350 ml, and spraying 80% sulphur at 1000 g/ha.

» For control of red spider mites, try using a Sulfur-based hydrocarbon oil mixture.

» Recent research has shown that red spider mites can be effectively treated with spray oil made with a paraffinic base. This oil is incorporated into the tea industry's mite control programme because it does not leave any residues.

» Predators of eggs and young insects include the stigmatid *Agistemus fleschneri*.

» Mites are prey for the coccinellids *Stethorus gilvifrons, Menochilus sexmaculatus*,

Cryptogonus bimaculatus, and *Jauravia quadrinotata*.

» Spider mite eggs being devoured by a predatory mite (*Phytoseilus* sp.).

» *Verticillium lecanii*, an entomopathogenic fungus, spore suspension @ 1.5 kg/ha applied in the evening when humidity is higher.

» It was discovered that *Paecilomyces fumosoroseus* ('Mycomite') works well.

» Use of the Biogreen-5 Technology (*Trichoderma viride, Pseudomonas fluorescens, Beauveria bassiana, Metarrhizium anisopliae* and *Bacillus thuringiensis* based consortial bioformulation).

DISEASES

Twig dieback and stem canker, *Macrophoma theicola*

Symptoms

» Yellowing and wilting of leaves occur suddenly on damaged branches.

» Most branch ends perish.

» The bark and the stem acquire grey blotches, and subsequently sunken patches (cankers) emerge, girdling the stem.

» The plant tissue above the stem canker gradually loses strength and eventually dies.

» In hot and dry conditions, the signs of damaged plants become more noticeable.

Favorable conditions

» Spreading during wet weather and maturing in dry weather.

» Extremely humid (relative humidity over 85 percent).

» Stormy conditions with a strong wind.

» During the hail storm

» In order to infect a plant, the fungus must first obtain access, and this typically requires a wound.

Survival and spread

» Manage for a few weeks on the pruned branches that were left in the field.

» Splashes of rain can disperse spores, which can then

Management

- Maintain a thriving tea garden.
- Plant in an acidic, well-drained soil, protect the roots, and fertilise regularly for optimal growth.
- Pruning diseased branches several inches below the cankered patches is an effective way to remove them.
- Using a mix of one-part household bleach to nine parts water, disinfect the pruning shears between each usage.
- New leaf scars can be kept free from infection by applying fungicides like Thiophanate- Methyl and Copper salts of fatty acids during wet and regular leaf drop times.

Flower blight, *Ciborinia camelliae*

Symptoms

- Resulting in a browning of the blooms.
- Early spring is when flower blight is most visible, typically because of the abundance of moisture at that time.
- The first signs appear as little brown patches of erratic shape on the flower petals.
- These dots rapidly spread across the flower's surface.
- Inside of twenty-four to forty-eight hours, the entire flower wilts and falls to the ground.
- The plant as a whole is unaffected, only the blossoms.

Spread

- This fungal disease spreads via air currents, eventually arriving on camellia blossoms and turning them brown and mushy

Management

- The most effective method of management is sanitary conditions. Take off the diseased flowers' stems and throw them away. All of the dead flowers, leaves, and other plant matter should be raked up and disposed of.
- Put new mulch around the base of the plant. The soil is where this fungus is able to live.

- Fungus spores can be carried by the wind for up to 1.6 kilometres.
- Control is most effective when sprayed to other tea plants in the area.
- Late December/early January is a good time to apply soil drenches around the plant, such as PCNB, to assist lessen the severity of disease.
- Mancozeb and Triadimefon are two fungicides that ought to be sprayed on the blossoms.

Root rot, *Phytophthora cinnamomi*

Symptoms

- The initial signs are a general yellowing of the leaves, stunted growth, and a general wilting of the plant.
- Discolored and missing tiny feeder roots are telltale signs of a root infection.
- Roots that have been infected range in colour from reddish brown to a deep chocolate brown (healthy roots are white).
- A plant might die suddenly, or it can slowly decrease over the course of several years.
- While all popular tea types are vulnerable to root rot, all Sasanqua camelliae cultivars are immune.

Favorable conditions

- Soils that are warm and poorly drained are ideal conditions for the development of many diseases.
- The fungus grows rapidly in humid, warm soils that have inadequate drainage.

Management

- Incorporation of tea plant varieties grafted onto Sasanqua (*Sasanqua camelliae)* resistant rootstocks.
- Look for plants that don't have any yellowing or wilting on their leaves and are otherwise in good health.
- Planting sites should always have good drainage.
- Using raised beds can help with drainage in already developed regions.
- Only prophylactic use of fungicides like Etridiazole and Mefenoxam is likely to be successful, and these fungicides require frequent reapplications.

Blister blight, *Exobasidium vexans, E. camelliae*

Symptoms

- Affect the young, harvestable leaves of the tea plant, leading in a catastrophic loss of production.
- Small, light or pinkish translucent dots emerge on fragile leaves just after the commencement of monsoon, after the first rainfall.
- The tender stem is impacted as well.
- Most instances of leaf galls are noticed during the period of rapid growth that occurs in the spring.
- The new growth and leaves become abnormally big, thick, and meaty.
- The damaged areas go from a pale green to a near-white or pink hue.
- Later, a white mass of spores is revealed as the galls burst on the underside of the leaves.
- Over time, the galls harden and turn dark.
- Rarely do plants sustain significant injury.

Favorable conditions

- In the monsoon months, when it's cloudy and the leaves stay damp for 11–13 straight hours, the humidity is typically above 60%, and the temperature is typically between 17 and 22 °C.

Survival and spread

- Because of the hospitable climate in hilly areas of tea gardens, the pathogen is able to remain dormant in tea plantations all year long.
- An outbreak can spread like wildfire if a rainy period is followed by four or five days of bright sunshine.

Management

- Pruning at the appropriate time, reducing shade, and keeping weeds at bay can all help minimise the spread of this disease.
- Galls should be removed and destroyed when they are young to prevent the bottom leaf surfaces from turning white and releasing spores. Collect and dispose of the leaves that have accumulated.
- Don't water the leaves too much or they'll fall off. Galls are more likely to

form when the weather is warm, wet, and overcast.

- Before an infection sets in, spray on some 0.25% Mancozeb, 0.05% Calixin, 0.1% Baycor, or 0.1% Bayleton. Spraying should begin at bud break and continue at 7-14 day intervals until June 1st. Use a 1% Bordeaux combination or 0.2% copper oxychloride solution as needed.
- To prevent blister blight in tea, spray Hexaconazole (1 ml/L of water) or Tridemorph (2 ml/L of water) every 15 days, then apply Copper oxychloride (3 ml/L of water) every 7 days.

Grey blight, *Pestalotia theae/Pestalotiopsis theae*

Losses

- Disease has caused a 17% drop in agricultural yield.

Symptoms

- This pathogen attacks mature leaves, new growth, and bare stems.
- Patches first develop on immature leaves as tiny, round, yellow-green spots.
- In many cases, the patches have a thin, yellow border.
- Defoliation occurs when the spots dry off and turn brown or grey, revealing concentric rings punctuated by tiny black dots.
- When a young shoot becomes infected, it dies back.
- Continual shear harvesting was a primary cause of shoot dieback.
- Dieback of young shoots directly leads to severe crop loss, whereas grey blight indirectly reduces bush health, which reduces production.
- From July to December, disease rates typically rise to their highest point of the year.

Favorable conditions

- When temperatures are high, the humidity is high, or the leaves stay moist for an extended length of time, the disease thrives.

Survival and spread

- The fungus spores are contained in the tiny, black dots on the sores.
- The spores can be spread from one plant or infected area to another via rain splash.

Management

- Reducing plant stress is a priority.
- Spacing your tea bushes out properly will help keep the humidity down and the leaves dry for a shorter amount of time.
- A hand-operated knapsack sprayer will be used to apply Mancozeb at a concentration of 0.3% (30 g in 10 L of water), Carbendazim, or Thiophanate Methyl at a concentration of 0.05% (5 g in 10 L of water), and the treatment schedule will be every 10–15 days.
- The use of Biogreen-5 resulted in the greatest decrease in grey blight (91.20 percent) (*Trichoderma viride, Pseudomonas fluorescens, Beauveria bassiana, Metarrhizium anisopliae* and *Bacillus thuringiensis* based consortial bioformulation).

Collar canker and dieback, *Phomopsis theae*

Symptoms

- The damaged tea bushes' leaves became smaller and yellowed before dropping off prematurely.
- Frequent periods of abundant blooming and fruiting are followed by periods of reduced canopy cover.
- A gradual dieback of the branches left the bushes naked.
- Cankers characterised by fissures and flaking of bark at the trunk's collar always accompanied foliar symptoms.
- Discolored tan, brown, or black patches covered the exposed cambium and wood.
- The collar area was affected by cankers that were uneven in shape and grew to a depth of several millimetres.
- Small, reddish, superficial perithecia were observed when the air was humid.
- There was also evidence of fibrous root development above the cankered region.

Favorable conditions

- Predisposing factors for disease development showed that soil composition plays a significant role.
- Disease rates were higher in gravelly soil.

Management

- Choose plants with a strong root structure.
- Pruning good wood away from the diseased wood and then treating the stubs with copper fungicide should solve the problem.
- Predisposing factors for the disease include deep planting, fertiliser treatment close to the plant collar, and mulching with raw coir pith close to the collar.
- Carbendazim, a systemic fungicide, was applied as a wound dressing and drenched into the soil, resulting in good control.
- Applications of biological control agents, such as *Trichoderma harzianum* and *Gliocladium virens*, to the soil around the bush and as a wound dressing, were more effective than either application alone.

Black root rot, *Rosellina arcuata*

Symptoms

- On the root's outer surface and around the collar, a black, fuzzy mycelium grows.
- Upon removing the bark from the bigger roots, the mycelium fungus is revealed as white, star-shaped sheets.
- On occasion, a black perithecia that resembles lead shot will appear on the collar.
- Mycelium spreads unrestrictedly through the top layer of soil and decaying vegetation.
- Mycelia surround the stem 7.5–10.0 cm below the soil's surface, where they suffocate the bark.
- Over the decaying area, the surrounding stem tissue will swell and form a ring.

Favorable conditions

- Predisposing factors for disease development showed that soil composition plays a significant role.
- Disease rates were higher in gravelly soil.

Management

- Choose plants with a strong root structure.
- Pruning good wood away from the diseased wood and then treating the

stubs with copper fungicide should solve the problem.

» Predisposing factors for the disease include deep planting, fertiliser treatment close to the plant collar, and mulching with raw coir pith close to the collar.

» Carbendazim, a systemic fungicide, was applied as a wound dressing and drenched into the soil, resulting in good control.

» Applications of biological control agents, such as *Trichoderma harzianum* and *Gliocladium virens*, to the soil around the bush and as a wound dressing, were more effective than either application alone.

Brown root, *Fomes noxius*

Symptoms

» Tea fields show signs of the disease in circular clusters.

» Fungi have mycelium that is a fawn colour.

» In the last stages of the disease, the diseased root decomposes and takes on a spongey consistency.

» Reticulations like honeycombs can be seen when such roots are cut in half lengthwise.

» Soil crusts thickly over the root's exposed surface.

» Rarely seen in their traditional bracket shape, fructifications are nonetheless possible.

Favorable conditions

» Lower-lying regions are more susceptible to disease.

» During the wet season, diseases tend to flourish due to the high humidity.

Spread

» Infection can be spread by root contact and alternate hosts after spores are transported by the wind and land on the stumps of shade trees (Coffee, Grevillea, Albizia and Erythrina).

Management

» Creating a perimeter around the diseased region by digging trenches 1.3 m deep and 45 cm wide.

» Burning and uprooting diseased shrubs.

- The use of Guatemala grass in soil restoration.
- The disease can be efficiently controlled by applying biocontrol agents (*Trichoderma harzianum, T. viride, T. hamatum, T. resei,* and *T. koningii*) at a rate of 200 g/plant.

Red root, *Poria hypolateritia*

Symptoms

- Foliage going yellow means something is wrong.
- The gradual or instantaneous dying of plant tissue.
- When a leaf dries down, it doesn't fall off the plant right away.
- If you pull the bush out of the ground, you'll see some discoloured red roots and some white mycelium.

Favorable conditions

- The rainy and humid weather is ideal for the spread of the disease.

Spread

- Mycelium in the soil is the major vector of infection because of the pathogen's microbial nature.
- Mycelium strands in the soil facilitate propagation.

Management

- Planting Guatemala grass to improve the soil.
- The diseased shrubs and neighbouring plants with yellowing leaves must be removed.
- Trees and stumps should be uprooted from the infected zone.
- By digging with a fork, remove any roots, both living and dead, that are as thick as a pencil.
- Burning the collected materials is the best option.
- Treat the bushes around the infected area as well.
- Employing biocontrol agents (*Trichoderma harzianum*) at a rate of 200 milligm per plant.

Algal leaf spot, *Cephaleurus parasiticus, C. mycoidea*

Symptoms

» Lesions appear on the leaves and tend to be round, elevated, and a purple or reddish brown colour.

» The first sign of infection is a brown discoloration at the stem's tip, cotyledon, or collar.

» After spreading to the petiole, the lesions eventually cause the leaves to droop and fall off.

» In the last stages of an infection, defoliation and dieback are typical.

Favorable conditions

» Tea plants are more susceptible to infection by algal leaf spot in conditions of poor soil drainage, nutritional imbalance, and exposure to relatively high temperature and humidity.

Survival and spread

» In the midst of the south-west monsoon, the disease is at its worst (June-September).

» The bacterium can only live in rotting and decomposing substances.

» It is the wind or rain that carries the "spores" to new locations. The alga might quickly infest other parts of the tree, including the fruit.

Management

» Reducing plant stress is a priority.

» Do not build in low-lying areas.

» Increase ventilation in the plant canopy to cut down on moisture retention on the leaves.

» All diseased and dead seedlings must be taken out of the nursery and destroyed.

» Copper hydroxide (Kocide) applied as a seed dressing and drenched into the soil is an efficient method of disease control.

» There was also success with soaking the soil in a Bordeaux combination diluted to 1 percent.

» The garden's drainage system needs to be upgraded.

References

Cranham, J. E. (1966). Tea pests and their control. Annual Review of Entomology, 11(1), 491-514.

Hazarika, L. K., Bhuyan, M., & Hazarika, B. N. (2009). Insect pests of tea and their management. Annual review of entomology, 54, 267-284.

Hazarika, L. K., Puzari, K. C., & Wahab, S. (2001). Biological control of tea pests. Biocontrol Potential and its Exploitation in Sustainable Agriculture: Volume 2: Insect Pests, 159-180.

Lehmann-Danzinger, H. (2000). Diseases and pests of tea: Overview and possibilities of integrated pest and disease management. Journal of Agriculture in the Tropics and Subtropics, 101(1), 13-38.

Kumhar, K. C., Babu, A., Arulmarianathan, J. P., Deka, B., Bordoloi, M., Rajbongshi, H., & Dey, P. (2020). Role of beneficial fungi in managing diseases and insect pests of tea plantation. Egyptian journal of biological Pest Control, 30(1), 1-9.

Rattan, P. S. (1992). Pest and disease control in Africa. In Tea: cultivation to consumption (pp. 331-352). Dordrecht: Springer Netherlands.

22 Turmeric

INSECT PESTS

Rhizome fly, *Mimegralla coeruleifrons*

Damage symptoms

- The maggot causes the holes in the turmeric plant stems by boring into them.
- To add insult to injury, maggots' presence causes decay in the rhizome.

Identification of pest

- The adult form of a fly is quite sizable, with a long, leggy body and a slim waist. The transparent wings have ashy-colored patches, but the body is completely black. Flies can spread their wings to a size between 13 and 15 mm.

Management

- Plant only from wholesome seed stock.
- Remove and kill the afflected vegetation.
- Application of Diazinon spray.
- 20% Phorate (about 20 kg per hectare) is being sprayed at the plants' roots.
- Treatment of ginger seeds with Chlorpyriphos 20 EC at 2 ml/L of water and drenching with the same dose of Chlorpyriphos, and application of Neem cake at 2 tons/ha at the time of planting.

Shoot borer, *Dicchocrocis* (*Conogethes*) *punctiferalis*

Damage symptoms

» Eggs are laid on leaves and other tender plant parts by the moths.

» Yellowing and drying of leaves on infested pseudo-stems is caused by larvae that dig into the plant tissue and feed on interior tissues, leading in the creation of dead hearts.

» A bore-hole on the pseudo-stem, from which frass is extruded, and a withered, yellow core shoot are telltale signs of insect infestation.

Pest identification

» Eggs are pink, oblong, and spherical.

» Adult larvae are hardly hairy and a pale brown colour.

» A medium-sized moth, the adult has a wingspan of around 20 mm and is orange- yellow with tiny black dots.

Favorable conditions

» Conditions: 30–33 °C with 60–90% relative humidity.

Management

» No pests may infest the rhizomes from which the turmeric seeds will grow.

» Don't let the water sit there doing nothing. It is essential that they be drained correctly.

» The installation of light traps is required for moth control.

» Take out the affected branches and throw away the caterpillars as well.

» Every two weeks, spray a 0.5% Neem oil solution.

» One kilogramme of Carbaryl 50WP diluted in five hundred litres of water, to be sprayed.

» During the months of July through October, a spraying schedule of 0.04% Endosulfan or 0.1% Malathion every 30 days is implemented.

» Parasitoids of the hymenopteran order (*Hexamermis* sp., *Bracon brevicornis*) have been found to be effective at reducing populations of this pest.

Leaf roller, *Udaspes folus*

Damage symptoms

- To begin feeding, newly hatched larvae scrape the leaves.
- Caterpillars later cut and fold leaves, retreating inside of them to live and feed.
- Leaves coil up in a longitudinal direction.
- The plant's leaves have been removed entirely.

Pest identification

- Pinkish, flat, oval eggs.
- Larvae that have reached maturity are a dark green colour with a black head and a narrowed neck.
- The pupa has a long, cylindrical shape and is a watery green tint. Its frontal projection is long and conical. The pupa's proboscis stands out most among its other features.
- The adult has a medium frame and brownish black wings with eight white dots on the forewings and one huge patch on the hindwing.

Favorable conditions

- Ranging from 26 to 35 °C.
- Humidity can be anything from 41 percent to 100 percent.

Management

- Regular and thorough cleaning and disinfection.
- Spraying with 0.2% Carbaryl aids in the control of the infestation when it becomes severe.
- Use a 0.05% solution of dimethoate or phosphamidon and spray it over the area.

Rhizome scale, *Aspidiella hartii*

Damage symptoms

- Scales cause problems for rhizomes both in the field and in storage.
- Rhizomes become withered and dry when they are badly infested with adult (female) scales because they feed on sap.

- When an infestation first begins, the white scales can be found spread across the rhizomes; later, they cluster around the buds.
- If the infestation is bad enough, the rhizome and the buds will begin to wither, and eventually the whole rhizome will dry up.
- Pests in storage cause the buds and rhizomes to shrivel up.

Pest identification

- The shape of a long egg.
- Creatures that crawl and are yellow in colour.
- The females' scales are encrustations on the rhizomes, and they are round, around 1mm in diameter, and a light brown to grey colour. The male has a bright orange body, clear wings, and a well-defined skull, chest, and abdomen.

Management

- Pick infested-free rhizomes to plant as seed.
- Damaged or infested leaves and rhizomes should be gathered for disposal.
- Seed rhizomes should be soaked in Malathion or Dimethoate at a 0.05% concentration for 30 minutes before being stored or planted.
- Phosphomidon (at a concentration of 0.05%) sprayed in the field is effective in preventing the pest's reproduction.
- After harvesting, before planting or storing, turmeric rhizomes are submerged in a solution of Quinalphos (0.025%) or Fenitrothion (0.05%) for 5 minutes.

Thrips, *Panchastotrhips indicus*

Damage symptoms

- Thrips feed off the plant sap and cause damage to the undersides of leaves.
- They are particularly destructive to tender plant parts like young leaves and branches.
- The leaves start to curl up, turn white, and eventually dry out.
- Young leaves wilt and dry out when there is a severe infestation.

Pest identification

- Adult thrips are very tiny insects that can be either yellow or black in colour. Their abdomens are also very lengthy. The adults of these insects may have wings, but they are rarely able to fly.

Favorable conditions

» Seasonal heat and humidity.

Management

» Spray with either 0.1% Carbaryl, 0.3% Dimethoate, 0.3% Methyl demeton, or 0.3% Phosphamidon.

White grubs, *Holotrichia* spp.

Damage symptoms

» Both the adult and the larval stages of grubs are harmful to plants.

» As they continue to thrive in the soil, grubs consume the plant's nourishing roots, leaving only the tap root.

» They can also harm a plant by burrowing through its pseudo stem near the ground.

» A plant that has been invaded by grubs will lose its green colour, its leaves and branches will fall off, and the entire plant will eventually wither and die.

» As it eventually fades off, it leaves behind uneven crop development.

» Fungus gnats, sometimes known as root grubs, can occasionally cause shoot yellowing and wilting by feeding on young, sensitive rhizomes, roots, and the base of pseudo stems.

Pest identification

» White and fairly round, eggs are a staple in any kitchen.

» Larvae have a globular head and an elongated, dorsoventrally flattened body, forming a C shape and moving slowly. Translucent and white, the immature grubs are just 5 mm in length.

» An adult *Holotrichia* spp. measures in at 18-20 mm in length and 7-9 mm in width.

Favorable conditions

» Seasonal heat and humidity.

Management

» Due to the pest's appearance in the root zone in dry soil conditions, it is imperative that appropriate irrigation be provided.

» When adult beetles are seen munching on your plants, collect them and kill them.
» You can catch adult insects by using light traps.
» The endemic regions should implement crop rotation.
» Near the root zone, use either Lindane 1.3D 125 kg/ha or Carbofuran3G 10 kg/ ha; repeat the treatment 30 days later.

DISEASES

Rhizome rot, *Pythium graminicolum*

Disease symptoms

» Beginning at the pseudo-collar, stem's the infection spreads both laterally and vertically.
» Soft rot occurs when the collar area of the damaged pseudo-stem becomes waterlogged, and the rotting process extends down into the rhizome.
» Later on, root infection becomes apparent.
» Symptoms on the leaves start as a slight yellowing at the leaf tips and work their way up the leaf blades.
» The disease's earliest stages are characterised by a partial greening of the leaves' interiors and a gradual yellowing of their peripheries.
» Subsequently, the yellowing progresses upwards through the plant's leaves from the lower region, and is followed by the drooping, withering, and drying of the pseudo-stems.
» Dying plants often have rotting rhizomes and roots.

Favorable conditions

» The infection is most lethal to the sprouts while they are young. Rhizome rot is made worse by nematode infestation.
» Conditions that promote the spread of the disease include temperatures exceeding 30 °C and very wet soil.
» Poor field drainage causes water accumulation, which exacerbates the condition.

Survival and spread

- This disease spreads through the soil. As soil moisture increases due to the south west monsoon, the fungus flourishes.
- The fungus can live in one of two ways: diseased rhizomes that are stored for sowing, and in the soil, where it overwinters in dormant structures like chlamydospores and oospores.

Management

- Spreading of disease-free seeds.
- After soaking in a solution of 3 gm of Mancozeb, 1 gramme of Carbendazim, 3 gm of Ridomil MZ, and 2.5 gm of Metalaxyl for an hour in 1 litre of water, the seed rhizomes should be dried in the shade for a day before being planted.
- Apply a mixture of 1 gramme of Mancozeb per litre of water and 1 millilitre of Carbendazim to the crop.
- *Peudomonas fluorescens* and *Trichoderma viride* in a consortial formulation at 4 g/ kg for seed treatment, plus 2.5 kg/ha of soil applied as a combination of a base and top dressing.

Leaf spot, *Colletotrichum capsici*

Disease symptoms

- The disease manifests as brown patches with black concentric rings of varying diameters on the upper surface of the young leaves.
- The centres of the dots are either white or grey, and their shapes are erratic.
- Spots can eventually join together to form larger, uneven patches that can spread throughout the entire leaf.
- Fruiting structures, resembling pinheads, can be found at the centre of most places.
- When attacked severely, leaves shrivel, dry up, and take on a burned look.

Favorable conditions

- Temperature of 25 °C, high soil moisture, and damp leaves.

Survival and spread

- Plant waste acts as a reservoir for the disease that is spread through the soil.

- Disease is spread through splashing rain during brief downpours. Curcumin is highly susceptible to the disease when cultivated in unprotected environments.

Management

- Sanitization procedures should be used in the field.
- The accumulation of inoculum can be lowered by rotating cereal and legume crops.
- Uproot and destroy any plants that have an infected leaf by plucking it off with your fingers.
- For less soil splashing, use adequate green mulching.
- Bordeaux spray, which is a 5:5:50 blend of wine, brandy, and water.
- Zineb (Dithane Z-78) spraying at a rate of 1.5 kg in 500 litres of water per hectare.
- Use a spray containing either Mancozeb (0.25%) or Copper oxychloride (0.25%) immediately after the onset of symptoms and again 15 days later.

Leaf blotch, *Taphrina maculans*

Disease symptoms

- The spots themselves are tiny and either oval or primarily rectangular in shape.
- Spots develop on both the upper and lower sides of infected leaves, however the spots are more prevalent on the upper side.
- They're lined up neatly in columns next to the arteries and veins.
- Spots spontaneously join together to develop lumpy sores.
- Discolorations start off as light yellow and eventually turn a dirty shade of that colour.
- The plants become burned and produce fewer rhizomes when the damage is severe.

Survival and spread

- Spread through soil and seed and able to overwinter in the soil on infected plant matter.

Favorable conditions

- The soil was very moist, the air was warm (about 25 °C), and the leaves were damp.

Management

- For less soil splashing, use adequate green mulching.
- Sanitization procedures should be used in the field.
- The accumulation of inoculum can be lowered by rotating cereal and legume crops.
- Spray made from garlic extract.
- Spraying with 0.25 percent Mancozeb or 0.25 percent Copper oxychloride as soon as symptoms develop, and again after 15 days, should help.
- Application of Propiconazole 0.1% as a foliar spray should be done 45 and 90 days following planting.

Dry rot, *Rhizoctonia bataticola*

Disease symptoms

- Dry rot of rhizomes, also known as root rot and rhizome rot, is a common symptom of this disease beginning in October.
- The affected rhizomes start out looking soft and shrunken, but then dry up and harden as time goes on.
- Normal signs of crop maturity, such as yellowing and drying of leaf in October and November, would be indistinguishable from those of disease-affected clumps.
- The diseased areas of rhizomes usually look dark brown when cut open.

Favorable conditions

- Soil temperatures of 35 °C, moisture levels of 15 to 20 percent, and alluvial or sandy soils are ideal for the disease's spread.

Survival and spread

- The pathogen is a saprophytic organism that may survive on organic substances in the soil for an extended period of time.
- Infestation begins in weak plants.

Management

- Sanitization procedures should be used in the field.
- Reduce the accumulation of inoculum by rotating cereal and legume crops.
- Dry root rot can be effectively managed with the help of *Trichoderma harzianum* and Neem cake.

Bacterial wilt, *Ralstonia solanacearum*

Disease symptoms

- The typical symptoms include rapid wilting and death of the entire plant without any yellowing or staining of leaves.
- When one branch dies, they all die.
- A wilted plant's pith will look black and wet when you cut it in half vertically.
- When the stem is pressed, a grey, slimy substance emerges.
- Pith degradation in the later stages of the disease can lead to significant hollowing of the stem.

Favorable conditions

- This bacterium is most harmful in warm, wet soils with temperatures over 24 °C.
- Infectious diseases thrive in warm, wet soil.

Management

- *Bacillus subtilis* and *Pseudomonas putida*-enhanced agricultural yard waste.

References

Shanmugam, P. S., Indhumathi, K., Sangeetha, M., & Tamilselvan, N. (2015). Evaluation of different pest management modules against major insect pests and diseases of turmeric. Current Biotica, 9(1), 17-24.

Muthomi, J., Fulano, A. M., Wagacha, J. M., & Mwang'ombe, A. W. (2017). Management of snap bean insect pests and diseases by use of antagonistic fungi and plant extracts. *Sustainable Agriculture Research*, *6*(3), 52-63.

Satyagopal, K., Sushil, S. N., Jeyakumar, P., Shankar, G., Sharma, O. P., Sain, S. K., ... & Sreenivas, A. G. (2014). AESA based IPM package for turmeric. GoI.

Rao, T. N. (1995). Diseases of turmeric (Curcuma longa L.) and their management. Journal of Spices and Aromatic Crops, 4(1), 49-56.

Nair, K. P., & Nair, K. P. (2019). Turmeric Entomology. Turmeric (Curcuma longa L.) and Ginger (Zingiber officinale Rosc.)-World's Invaluable Medicinal Spices: The Agronomy and Economy of Turmeric and Ginger, 125-144.

Koya, K. A., Devasahayam, S., & Kumar, T. P. (1991). Insect pests of ginger (Zingiber officinale Rosc.) and turmeric (Curcuma longa Linn.) in India. Journal of Plantation Crops, 19(1), 1-13.

Nair, K. P., & Nair, K. P. (2019). Diseases of Turmeric. Turmeric (Curcuma longa L.) and Ginger (Zingiber officinale Rosc.)-World's Invaluable Medicinal Spices: The Agronomy and Economy of Turmeric and Ginger, 151-171.

Devasahayam, S., Jacob, T. K., Eapen, S. J., & Senthil Kumar, C. M. (2022). Pests and Their Management in Spices: (Black Pepper, Cardamom, Ginger, Turmeric, Cinnamon, Clove, Nutmeg and Vanilla). Trends in Horticultural Entomology, 1331-1364.

23

Vanilla

INSECT PESTS

Scale insect, *Conchaspis angraeci*

Damage symptoms

- Mostly the branches and leaves showed signs of it.
- These dark spots can be a nuisance, especially if they appear repeatedly on the stems, leaves, or flowers and refuse to move.
- This armoured scale causes chlorotic blotches on leaves by sucking the plant's sap, which then causes the leaves to dry out and, in the worst circumstances, kills the vine.
- The yellow spots on the stem and foliage are caused by the toxic saliva of the Conchaspis angraeci.

Pest identification

- Body is purple with a yellowish brown pygidium; eggs are light yellow. Scale cover is about round, white, with 6 to 8 ridges radiating from a central cone without exuviae.

Management

- Maintain a healthy environment for plant growth and give your plants the attention they need, including regular, sufficient watering.
- Pruning to open up your canopies.
- To keep ants at bay, you can apply a sticky substance (Tanglefoot) to tree trunks.

- Horticultural (narrow-range) oil sprayed at the right time and in sufficient quantity during the dormant season or soon after scale crawlers become active in late winter to early summer will effectively manage most species of scales.
- Apply a spray of insecticide, such as a soap, oil, or canola oil.
- Neonicotinoids (Acetamiprid, Dinotefuran, Imidacloprid, and Thiamethoxam) and Organophosphate (Acephate) are persistent, broad-spectrum insecticides that are applied.
- Protecting parasitic wasps (*Aphytis, Coccophagus, Encarsia*, and *Metaphycus* species) and predatory insects (including certain beetles, bugs, lacewings, and mites).

Weevils, *Sipalus* sp.

Damage symptoms

- The black weevil eats young, fragile stems and leaves by boring into them.
- Adult weevils and their grubs cause necrosis and decay in vanilla by feeding on the plant.
- Grubs dig into the vine's inner tissue while weevils feed on the plant's leaves and outside.
- Everything along the length of the vine rots, falls apart, and shrivels up.
- Adults primarily use their snouts to enter into shoots for a meal; within a day, the necrotic area causes the shoot tip or vine to perish.
- The weevil eats leaves by scraping away the epidermis and mesophyll and leaving behind a thin, transparent epidermis on the underside or top of the leaf.

Pest identification

- The adult weevil measures 8-10 mm in length and is either a bright or dark black colour with two wavy white cross stripes on the elytra. In adulthood, women tend to be considerably bigger than men.
- Early grubs are yellow, and mature grubs are white with a brown capsule on their heads.

Management

- Weevil adults should be collected and disposed of between the months of November and February.

Sucking bug, *Halymorpha picus*

Damage symptoms

- This insect pest feeds by sucking sap from leaves and branches.
- Sucking insect adults and juveniles dry out and decay young shoot tips and budding inflorescence.
- The nymphs and adults of the beetle drew the sap from the young shoots, young leaves, and young inflorescences of the vine.
- The depletion of the sap caused pinprick-like punctures, and ultimately necrosis and rotting became apparent.
- Within 3–5 days, the affected shoot tips and inflorescence markings rotted and fell off.
- Possibly 40% of financial failures can be attributed to you.

Management

- In the months of November through February, collect the egg mass while the nymphs are in their initial stage.
- To control the pest, spray the young shoot tips and emerging inflorescence with a 0.05% solution of either Endosulfan or Quinalphos.

Snail, *Achatina fulcia*

Damage symptoms

- Snails prey on young, sensitive shoot tips and leaves (causing them to turn white) and cause the most damage where there is moisture and shade in the plantations.
- Do damage to the plant's limbs, blossoms, and fruit.
- Everything goes down when darkness falls.

Management

- The process involved hand-selecting and destroying.
- Don't use mulch that will take too long to break down.
- Around the plant's base, scatter some diatomaceous earth, coffee grinds, or broken egg shells. This should slash at the pests as they scurry about your plant.

- Snails were drawn to a mixture of rice bran and jaggery (19.66%) and papaya fruit waste (16.46%).
- The highest death rates were found with Metaldehyde 2.5% pellets at 5 kg ha-1 (94.86%), Crystal salt at 25 kg ha-1 (90.35%), and Bleaching powder at 25 kg ha-1 (83.50%).
- Use the Sevin dust spray.
- Snailkill poison bait is effective for preventing insect infestations

DISEASES

Root and stem rot, *Fusarium oxysporum* f. sp. *vanillae*

Symptoms

- Deadening and browning of the roots beneath the ground.
- Root destruction starves plants to death by cutting off their supply of water and nutrients at the base of the plant.
- Dropping of delicate tips, leaf and stem yellowing, and stem shrivelling are all signs of nutrient deficiency.
- When a plant's water and food supply is cut off due to a failure of its underground root system, the plant will use its energy multiplying those roots.
- Roots that are too old to continue growing because of the newer ones crowding them out eventually perish.

Management

- Making use of disease-resistant cuttings that haven't been treated with any pathogens.
- The eradication and destruction of defective or otherwise unusable tissue.
- The plantation must be provided with sufficient drainage.
- Using a Benomyl/Mancozeb solution to soak cuttings for 20-30 minutes.
- Carbendazim 0.1% spraying and soaking.
- Implementing organic alterations that foster the development of hostile organisms.
- Utilization of either *Trichoderma lactis* and *Bacillus pantothenticus* or *Bacillus firmus* and *Pseudomonas fluorescens.*

» Twice-applied *T. harzianum*-enhanced compost to the soil.

» Rhizobial consortiums (IISR-147 and IISR-148) containing *T. harzianum* were applied to soil.

» Non-pathogenic *F. oxysporum* strains are commonly utilised as a biocontrol agent to combat the more dangerous pathogenic strains.

» The Sarapiqu 1 (S1) cultivar is resistant to both Fusarium and drought. The beans of the S2 variety are resistant to Fusarium and contain very high concentrations of the flavouring compound vanillin (a natural antibiotic).

Bean rot, *Phytophthora meadii*

Symptoms

» Causes decay in beans, plant parts, and soil.

» A telltale sign of black rot is the appearance of water-soaked sores on the fruit.

» More of the fruit and eventually the vine's branches will decay as the process progresses.

» Infected stems and branches eventually become a pale yellowish brown and die.

» Only young, easily replaced parts are affected by the disease and die.

» Bean rot begins at the bean tips, where it forms a water-soaked lesion that spreads slowly to the rest of the beans in the bunch and, in some cases, exhibits extensive mycelial development on the outside.

» In extreme circumstances, the entire bunch of beans will rot and fall off.

Favorable conditions

» Prolonged wet weather, poorly draining soil, excessive shadowing of plants, overcrowded plants, and a lack of weed control in the plantation are all conducive to the onset of disease.

» The monsoon season is when the disease is at its worst, especially in shady plantations.

Management

» Remove and dispose of infected plant components.

» Every week, spray a mixture of 1% Bordeaux and 0.1% Carbendazim.

- Use of *Trichoderma harzianum*-enhanced compost applied to the soil twice.
- Rhizobial consortiums (IISR-147 and IISR-148) containing *T. harzianum* were applied to soil.

Anthracnose, *Colletotrichum vanillae*

Symptoms

- Effects the vanilla plant as a whole, including the top of the stem, the leaves, the fruits, and the flowers.
- Leaf spot, stem rot, and bean spot or rot are all caused by this fungus.
- Leaves were found to have lesions of a round or oval shape, ranging in colour from reddish brown to brown.
- Black spots, which are the fruiting structure of the fungi, appear as lesions or other signs of disease on the affected part's surface.
- As a rule, rotting leaves were seen on the older ones.
- Beans initially showed water-soaked lesions, which gradually transformed into brownish to blackish brown lesions of varying diameters.
- The infection causes harm to the plant's leaves and stem, stunting the plant's development.
- It's also common for diseased fruits to ripen too quickly and fall off the plant, further decreasing its output.

Favorable conditions

- Excessive shade of plants and overcrowding promote the development of diseases.

Management

- Remove and burn infected plant components.
- Maintaining a healthy balance of humidity and sunshine requires routine tree pruning.
- Strengthen plant vitality (e.g. nutrition with organic matter).
- Spray a 0.1% solution of Bayleton, Benomyl, Prochloraz, Carbendazim, and Hexaconozole, as well as a 0.3% solution of Mancozeb or a 1.0% solution of the Bordeaux mixture.

Root rot, *Sclerotium rolfsii*

Symptoms

- The bean, the leaf, and the root are all affected.
- Symptomatic of this condition is the decay of the vanilla bean tip, which reveals a dense mycelial mat on the infected surface.
- Beans and leaves that had come into contact with the contaminated soil quickly began to rot.
- Bean rot might affect only a few beans or the entire bunch.
- The decay started at the bean tips and spread up the stalk.
- In extreme circumstances, a white fluffy fungal film spreads across the entire bunch, covering every bean.
- The collar area also rotted.
- First the root tips rot, then the entire root system, and finally the vines wilt and turn yellow.
- There may also be water damage at the base of the stem, which turns necrotic and kills the plant.

Favorable conditions

- When the monsoons start to pour down, root rot often appears.

Management

- Reduce mulch and dig up diseased roots.
- Copper oxychloride or PCNB at 0.2%, 2–3 litres per plant, should be drenched into the soil and sprayed onto the plants.
- Drenching the soil with 0.1% Carbendazim has proven to be helpful.
- Mixed applications of FYM/compost with *Trichoderma harzianum*, *T. lactae*, and *Bacillus pantothenticus*.

Dry rot, *Rhizoctonia solani*

Symptoms

- To put it simply, that's a major reason why vanilla gets dry root rot.
- Initially manifesting at the plant's collar, symptoms eventually spread downward, causing dry rot at the root.

- Roots, stems, and leaves all wither as a result.
- Later stages showed shredding and shrivelling of the affected area.
- Stems, roots, and leaves wither and turn yellow.
- The summer months saw a spike in the prevalence of several diseases.
- At a later stage of the disease, the stems and leaves will begin to dry out intermittently.

Favorable conditions

- Summer is when the disease is at its worst.

Management

- Do not overwater; rather, cut back on irrigation.
- Weed out and dispose of any infected plants, and then clean up the soil around the affected area.
- Before and after the monsoon, use dry leaves as mulch around the stem's base.
- When cultivating, take care not to damage the roots.
- If a plant's condition has deteriorated to the point where all of its leaves, beginning at the soil level and working their way up, are wilted, it should be thrown away.
- Soak the soil near the plant's root with a Bordeaux 5% combination.

Mosaic disease, *Cymbidium Mosaic Virus* (Potexvirus)

Symptoms

- The newest leaves are misshapen and spotted with yellow or white.
- As the leaves age, they develop tiny brown patches.
- The stems become discoloured in large, brown spots.
- Leaves fall off and stems perish after a while.
- It could take anything from a few months to a year for the entire plant to perish.

Management

- Only plant cuttings that have not been exposed to any diseases.
- Exterminate diseased vegetation by cutting it down and burning it.

References

Vanitha, K., Karuppuchamy, P., & Sivasubramanian, P. (2011). Pests of vanilla (Vanilla planifolia Andrews) and their natural enemies in Tamil Nadu, India. International Journal of Biodiversity and Conservation. 22.

Mahadeo, K., Palama, T. L., Côme, B., & Kodja, H. (2022). Vanilla: Culture, reproduction, phytochemistry, curing, pest, and diseases. In Orchids Phytochemistry, Biology and Horticulture: Fundamentals and Applications (pp. 329-340). Cham: Springer International Publishing.

Havkin-Frenkel, D., & Belanger, F. C. (Eds.). (2018). Handbook of vanilla science and technology. John Wiley & Sons.

Odoux, E., & Grisoni, M. (Eds.). (2010). Vanilla. CRC Press.

24 Yam Bean

INSECT PESTS

Rose beetle, *Adoretus versutus*

Damage symptoms

» The rose beetle is a pest that can do significant harm to nurseries and new yam bean fields.

» Because they lose their leaves so quickly, seedlings are easy prey for predators.

» Nurseries and newly planted seedlings should be attacked between the third and fourth month.

» Adults make feeding holes in the leaflets by perforating them from the centre out, taking care not to damage the ribs.

» The leaflets have been eaten away in several tiny places, giving them a skeleton-like look.

» There are more assaults near the leaflets' tips than they are at their roots.

» Adult rose bugs leave behind a ragged border of depressions around the regions they've devoured.

» In the wee hours of the night, rose beetles snack on the plant's leaves.

» An adult's diet can cause serious harm or death to plants as young as 18 months old.

Survival and spread

» The most typical method of spreading rose beetle to new areas is via transporting affected plant parts (particularly leaves) to those locations.

» It is also possible for the larvae in the contaminated nursery to be spread to new locations at an early stage by soil movement.

Management

» Since the larvae tend to emerge at night and travel across the soil surface to attack the host plant, it has been suggested that the adults be picked off by hand using a lantern.

» Putting up physical barriers (like coconut fronds or bamboo fences) around each plant, provided that these obstacles are at least as tall as the leaves.

DISEASES

Rust, *Phakopsora pachyrhizi*

Symptoms

» As the disease spreads, more and more little brown or brick-red patches show up on the leaves.

» These lesions result in the formation of pustules called uredinia, which often appear on the underside of the leaf and release numerous urediniospores.

» When a leaflet develops a lot of lesions, the affected area turns yellow and eventually the whole leaflet falls off.

» Plants with severe diseases may lose all of their leaves.

» Having less healthy leaf tissue means fewer and smaller seeds, which in turn means less harvest.

» Petioles, stems, and even cotyledons are susceptible to rust as well as leaves.

» Pustules can darken to black as they mature.

» The transformation of uredinia pustules into telia occurs due to the creation of a layer of teliospores within the pustules.

Survival and spread

» In the absence of a living host, the fungus can persist as teliospores (the teliospores' thick walls protect the fungus from the environment) (over seasoning).

» Inoculum can be obtained from a wide variety of hosts, as this disease can survive on a wide variety of hosts.

» The inoculum for the pathogen is dispersed through the air (urediniospores).

Management

- Possible protection against the ailment is a shift in when crops are planted and harvested.
- More space between rows and fewer plants per plot could speed up the drying of the canopy.
- Some research suggests that boosting disease resistance can be accomplished by adjusting soil fertility, namely potassium and phosphorus levels.
- Only the fungicide chlorothalonil has a protectant mode of action among the Chloronitrile group. If there is new growth or weathering, Chlorothalonil may need to be reapplied multiple times for it to be effective.
- Protectants like the Strobilurins are effective for up to 2 weeks following administration, but they must be used before infection begins.
- As a result of the triazole's fungicidal properties, the plant's fungus is killed and no pustules or spores are produced. For about a month following application, they begin to help shield young plants from harm.

References

Ameh, G. I., & Okezie, C. E. A. (2005). Pests and diseases of African yam bean, Sphenostylis stenocarpa (Hoechst. ex. A. Rich) harms. Bio-Research, 3(1), 14-20.

Afolabi, C. G., Ogunsanya, O. M., & Lawal, O. I. (2019). Evaluation of some African yam bean (Sphenostylis stenocarpa [Hochst. Ex A. Rich]) accessions for resistance to flower bud and pod rot diseases. Current Plant Biology, 20, 100126.

Reddy, P. P., & Reddy, P. P. (2015). Yam bean, Pachyrhizus erosus. Plant Protection in Tropical Root and Tuber Crops, 267-279.

BouraÃ, A., Sikirou, R., Zocli, B., Ezin, V., Adjanonhoun, A., & Hell, K. (2020). Pests and Diseases Associated with Yam Bean (Pachyrhizus erosus) Cultivation at Farmersâ€™ Levels in the Central Region of Benin. Biosaintifika: Journal of Biology & Biology Education, 12(2), 132-139.

25 Yam

INSECT PESTS

Scale, *Aspidiella hartii*

Damage symptoms

- To coat in a hard scale during storage; also occurs on the non-thriving plant parts of the stem and leaves.
- Plants that are being continuously sapped by an infestation will eventually wilt.
- Use of infected tubers as planting material spreads the problem.
- While in storage, this species can expand to completely cover tubers in the shape of tiny, more or less white scales.
- Tubers may shrivel and die from an infection, ruining their quality, usefulness, and marketability.
- Though no actual volume is lost, the germination process may be slowed or halted as a result.
- Low germination rates and diminished flavour in cooked goods are the results of a scale infestation.

Pest identification

- Scale insects, both adult and juvenile, are pale yellow.

Management

- Do not use contaminated soil.
- Because yams are a food source for the insect, they should not be planted immediately following ginger.

- Cassava or yam tubers are dipped in a 2% petroleum ether extract of cassava or yam bean seed oil containing 0.01% surfactant for 10 minutes before being dried in the shade.
- The occurrence of scale insects was reduced by 92.3% when the tubers were stored in sand.
- Soaking the tubers for 10 minutes in Monocrotophos (0.05%), 0.3% Aldrin, or a mixture of 5.6 ml Diazinon 20% EC and 425 ml of White oil in 13.5 L of water is recommended prior to planting.
- It is common for the aphelinid *Physcus comperi* and the encyrtid *Adelencyrtus moderatus* to parasitize white scale nymphs.
- Parasitoids like *Coccobius* (Physua) *comperi* and *Adelencyrtus moderatus* play an important role in keeping the yam scale insect population in line.

Greater yam beetle, *Heteroligus meles*

Damage symptoms

- Adult beetles eat the tubers, leaving holes in them as they tunnel through the earth to get to them.
- During storage, tubers deteriorate in both look and quality and are more likely to rot.
- The tuber spoils afterward.
- The adult beetles consume the seedlings, which might cause the plants' eventual demise.
- Tubers with holes in them are worth less because of that.
- Since a high infestation makes tubers more susceptible to rot organisms, it usually leads in a total economic loss.

Pest identification

- Insects of the genus *Heteroligus* are over 30 mm in length and have a dark brown to black appearance.
- Adult beetles range in length from 23 millimetres to 33 millimetres and are a dark brown to black colour.
- The bodies of larvae are a creamy white to grey colour, and their heads are a light brown.

Spread

» The adults travel to different farms and plots by means of aerial migration.

Management

» Putting in your seeds as late in the season as you can (June/July).

» To keep tabs on the adult beetles, you can set up light traps.

» Avoid growing host plants by rotating crops.

» The yam beetles are repelled by a mulch made from the leaves of the mosquito plant (*Ocimum viride*) or lemon grass (*Cymbopogon citratus*).

» Before you plant your yam setts, dust them with wood ash powder. It is also possible to soak the setts in a solution of two parts wood ash to five parts water for 20 minutes, then dry them in the shade before planting.

» Effective beetle control can be achieved by applying a combination of pirimiphos- methyl and deltamethrin once weekly for three weeks, beginning 12 weeks after planting.

» Carbofuran-treated soils produced the maximum yield of tubers (19.67 t / ha) and the fewest holes caused by potato beetles.

» The 'Tamenyo' cultivar of white Guinea yam supposedly has a far lower susceptibility to beetle damage.

Mealy bugs, *Ferrisia virgata, Planococcus citri, P. dioscoreae, Pseudococcus citriculus, Rhizoecus* sp., *Geococcus coffeae, Phenacoccus gossypi*

Damage symptoms

» Powdery nymph and adult food covers the surface of the tuberous plant.

» When mealy bugs are present, germination rates drop and food quality suffers.

» Develop a type of white powder towards the top of the tubers, especially while they are in storage.

» Sprouts can die from exposure to these chemicals, making tubers useless as a seed source.

» The yam tubers and roots are a favourite food for mealy bugs, which can spread to devastating levels, shrivelling stored tubers.

» High populations of yam mealy bugs cause the leaves to dry out and wither.

Management

- Only wholesome seeds and soil should be used.
- If the plant's upper portions are infested, you should spray them with Malathion or Malathion mixed with an Oil emulsion (such Triona or Albolineum).
- Storing seed tubers requires a combination of 100 parts sand to 1 part Carbaryl dust (10%).
- Dimethoate (in concentrations as low as 0.05%) can be utilised to prevent further dispersal.
- Twenty-five percent of the *F. virgata* samples have parasites from the species *Blepyrus insularis*. Parasitism by *Acerophagus* sp. is also documented in *F. virgata.*

Foliage beetles, *Crioceris livida, Lema armata*

Damage symptoms

- Both the larvae and the adults feed on the tender new growth of the plant.
- That's a euphemism for defoliation.
- Larvae will use their own waste as a covering.
- Major local harm can be done, especially in the crop's early months of development.

Pest identification

- The length of a larva is up to 1 centimetre, and it is thick and meaty, glossy, black, and covered with mucus.
- The adult stage is characterised by a reddish brown to black body and elytra with dark brown to black markings.

Management

- The larvae should be hand-picked and killed.
- Use pesticides such as carbaryl, dichlorvos, or endosulfan to spray the area.
- The tachinid parasitoid *Pseudalsomyia* sp. frequently attacks *Lema lacordairei.* Parasite prevalence in the field is 7.9 percent.

Termites, *Coptotermes* sp., *Amitermes evuncifer, Protermes minutus*

Damage symptoms

- » It is possible that the termites infested the area before they made their way into the storage facility.
- » When colonies only use a single gallery to penetrate, the resulting damage can be hard to spot before it's too late.

Management

- » The production of tubers was dramatically (P) enhanced when planted on ridges, and the number of termites was drastically decreased.
- » The addition of N, P, K, and Mg (12: 12: 17: 2) successfully reduced the population of termites in the yam and increased the output.
- » Yam tubers become more infested with termites since they weren't harvested in a timely manner.

DISEASES

Anthracnose, *Glomerella cingulata* (*Colletotrichum gloeosporoides*)

Symptoms

- » Young leaves oftentimes get incredibly tiny brown patches.
- » As the leaves mature, these spots may expand and have thin, yellow borders.
- » Sometimes the spots on the leaves will join together to form larger irregular blotches, and the centres of these spots may fall out, creating a "shot hole" appearance.
- » Normal leaf drop occurs when an infection occurs.
- » Any diseased veins will turn dark on the underside of the leaf.
- » Similar signs, including a superficial but strong blackening, have been observed in stems.
- » Some of the lowest leaves on immature plants might make it through dieback, but more often than not, the whole vine dies and there are no or very few fruits produced.
- » "Dead skin," an orange-brown rot caused by fungi, affects tubers.

Favorable conditions

- Wet and humid weather is ideal for the spread of infection.
- Epidemics tend to spread more quickly during wetter, longer periods.

Survival and spread

- Acervuli, the dormant form of a disease, survive the winter in plant detritus.
- Perhaps the most common way in which diseases are spread to new crops is through the use of infected planting material.
- The conidia of fungi are disseminated by the splash of raindrops.

Management

- Eliminating rogues of diseased plants helps limit the spread of disease.
- Damage might be postponed by planting early.
- Intercropping and/or crop rotations with plants like maize.
- Using maize as a "barrier crop"
- Supporting plants with stakes early on is essential.
- Copper fungicides can be applied to the vines.
- Disease severity is lessened by spraying (0.2%) Mancozeb and (0.25%) Carbendazim.
- It was determined that *Trichoderma harzianum, Trichoderma viride*, and *Pseudomonas fluorescens* were effective.
- Planting yam types with natural resistance to anthracnose, such as TDA 291 or TDA 297, is the most successful approach of disease management.
- Treatment of soil and tubers with 50 g of Trichoderma mixed in fresh cow dung slurry per kilogramme tuber before to planting, in addition to foliar application, has been shown to significantly increase yields of the treated crop.

Tuber rot, *Fusarium oxysporum, F. solani, Rhizopus nodosus, Botryodiplodia theobromae*

Symptoms

- Brownish dry rot appears in infected tissues.
- The centre of the affected area will have brown dry rot caused by *Fusarium*, and the surrounding area may be blackish due to extensive damage and secondary bacterial infection.

- *D. esculenta* is especially vulnerable to the soft rot caused by *Rhizopus nodosus.*
- Tubers lose as much as 75 percent of their original volume as their skin dries out and wrinkles.
- The tubers turn brown at the cut and a viscous liquid of an unusual hue drains from the wounds.
- As the infection spreads through the tuber, only the vascular tissue is spared.
- As the diseased tissues within the tuber dry out, the tuber eventually takes on a mummified appearance.
- Wet rot, soft rot, and brown dry rot are all forms of tuber rot caused by *Botryodiplodia theobromae.*
- As a result, the tuber rots away in the ground and, especially, while it is stored.
- Pathogens that cause tuber rot gain admission to tubers more easily when they are infested with nematodes.

Management

- Planting material, such as cut tubers, should be treated with a Bordeaux mixture or coated with lime, chalk, or wood ash before being utilised in the garden.
- Systemic fungicides like Captan, Benomyl, and Thiabendazole have been shown to be effective when dipped into the tubers.
- Disease rates can be decreased through crop rotation.
- Before being stored in pits or barns, white yam tubers are treated with chemical fungicides (Mancozeb Metalaxyl WP) applied to their surfaces.
- Storage of tubers after *Bacillus subtilis* application.
- *Trichoderma viride* soil application (1 kg/ha) and a set treatment (30 g/L for 15 minutes).

Neck rot, *Corticium rolfsii*

Symptoms

- Disease symptoms include concentric rings of round leaf patches of varied diameters.
- When the fungus reaches maturity, sclerotia develop at the epicentre of the leaf spots.

- Necrosis can cause the lesion's centre to slip out, causing the lesion to swell and finally fuse with its surroundings.
- These round leaf spots are easily observable in the field and in nurseries, where they cause not only spots but entire blight of growing yam.
- The diseased yam vine also develops sclerotia at its root system.
- The petiole, which follows the leaf, is sometimes impacted.
- The plant's aerial parts wither and the lower stem turns black.
- Plant death is more likely to result from a neck infection.
- If you dig up a tuber, you'll notice that the tissues at its neck (the area where the tuber meets the stem) are in a complete state of disarray.
- More than a 50% drop in output is possible under extreme conditions.

Survival and spread

- The pathogen was discovered in soil, plant litter, and numerous weeds, including *Aclipha ciliata, Chromolaena odorata, Euphobia heterophilla, Ipomea triloba,* and *Commelina erecta.*
- Transfer of sufficient numbers of sclerotia during an infection can occur through mere rain splashing.

Management

- Before planting, early germination and depletion of the sclerotia are encouraged by early ploughing and rigorous disking.
- Soil solarization during the warm season raises soil temperatures, which, when maintained for an adequate amount of time (4-6 hours at 50 °C and 3 hours at 55°C), kill sclerotia.
- Rotating crops with ones that aren't hosts or that have resistant cultivars is a good strategy.
- The leaves of the vines can be kept off the ground with the use of supports, which can be trained.
- Lower rates of Sclerotium leaf blight were seen on yam fields that had recently undergone fire.
- Neck rot of yam was greatly decreased by a 0.1 g/ml concentration of ash from the rachis of plantain (*Musa* sp.) inflorescences.
- *Trichoderma harzianum, T. viride, Bacillus subtilis,* and *Gliocladium virens* are

four antagonistic fungi that have been found to be effective against *S. rolfsii*.

- *D. esculenta*, a resistant species, is utilised.

Mosaic disease, *Yam Mosaic Potyvirus*

Symptoms

- Yellow and green patterns (called mosaics) between the veins, or narrow green strips bordering the veins, are a common sign in diseased leaves (called vein banding).
- The leaves become severely chlorotic, with symptoms including green vein-banding, curling, mottling, green spotting, flecking, mosaic, shoe-stringing, interveinal chlorosis, stunting, and distortion.
- It causes a decline in photosynthetic capacity, which in turn reduces tuber yield and quality and can even cause plant death.
- Shoe-string symptoms, when the plant as a whole is stunted and the leaves become long and narrow, indicate a severe case of the disease.
- Some little tubers with a lower starch content could grow on the plant.
- However, even if a plant appears to have recovered from a virus infection shortly after the first symptom, the virus may persist in the plant and weaken its vitality.
- Mosaics, distortions, and, in extreme cases, strap-like leaves, low growth, and smaller tubers are all symptoms of a plant with a disease.

Transmission

- Insect vectors include the non-persistent aphids *Aphis gossypii*, *Aphis craccivora*, *Rhopalosiphum maidis*, and *Toxoptera citricidus*.
- The infected tubers or setts are then utilised to grow more of the same plant, which speeds up the disease's spread.

Management

- Whether grown from mini-tubers in a vector-free environment or from tissue culture, disease-free plantlets have multiple points of origin.
- Make sure fields are weed-free.
- Clean up and dispose of crop remnants.

References

Akoroda, M. O. (1993). Yams: Dioscorea spp. In Genetic improvement of vegetable crops (pp. 717-733). Pergamon.

Amusa, N. A., Adigbite, A. A., Muhammed, S., & Baiyewu, R. A. (2003). Yam diseases and its management in Nigeria. African Journal of Biotechnology, 2(12), 497-502.

Atiri, G. I., Winter, S., & Alabi, O. J. (2003). Yam. In virus and virus-like diseases of Major Crops in Developing Countries (pp. 249-268). Dordrecht: Springer Netherlands.

Reddy, P. P., & Reddy, P. P. (2015). Yams, Dioscorea spp. Plant Protection in Tropical Root and Tuber Crops, 193-233.

Jayaprakas, C. A., & Harish, E. R. (2022). Pests and Their Management in Minor Tuber Crops: (Chinese Potato, Yams, Taro, Elephant Foot Yam and Tannia). Trends in Horticultural Entomology, 1109-1137.

FSC
www.fsc.org
MIX
Papier | Fördert gute Waldnutzung
FSC® C083411

Kolibri 360 GmbH
Ferdinand-Jühlke-Straße 7
99095 Erfurt, Deutschland
produktsicherheit@kolibri360.de